Sustainable Civil Infrastructures

Sustainable Civil Infrastructures (SUCI) is a series of peer-reviewed books and proceedings based on the best studies on emerging research from all fields related to sustainable infrastructures and aiming at improving our well-being and day-to-day lives. The infrastructures we are building today will shape our lives tomorrow. The complex and diverse nature of the impacts due to weather extremes on transportation and civil infrastructures can be seen in our roadways, bridges, and buildings. Extreme summer temperatures, droughts, flash floods, and rising numbers of freeze-thaw cycles pose challenges for civil infrastructure and can endanger public safety. We constantly hear how civil infrastructures need constant attention, preservation, and upgrading. Such improvements and developments would obviously benefit from our desired book series that provide sustainable engineering materials and designs. The economic impact is huge and much research has been conducted worldwide. The future holds many opportunities, not only for researchers in a given country, but also for the worldwide field engineers who apply and implement these technologies. We believe that no approach can succeed if it does not unite the efforts of various engineering disciplines from all over the world under one umbrella to offer a beacon of modern solutions to the global infrastructure. Experts from the various engineering disciplines around the globe will participate in this series, including: Geotechnical, Geological, Geoscience, Petroleum, Structural, Transportation, Bridge, Infrastructure, Energy, Architectural, Chemical and Materials, and other related Engineering disciplines.

SUCI series is now indexed in SCOPUS and EI Compendex.

More information about this series at http://www.springer.com/series/15140

J. James Yang · Wen-Chieh Cheng ·
Shuying Wang
Editors

Advanced Tunneling Techniques and Information Modeling of Underground Infrastructure

Proceedings of the 6th GeoChina International Conference on Civil & Transportation Infrastructures: From Engineering to Smart & Green Life Cycle Solutions – Nanchang, China, 2021

Springer

Editors
J. James Yang
School of Civil Engineering,
College of Engineering
University of Georgia
Athens, GA, USA

Wen-Chieh Cheng
School of Civil Engineering
Xi'an University of Architecture
and Technology
Xi'an, China

Shuying Wang
School of Civil Engineering
Central South University
Changsha, China

ISSN 2366-3405 ISSN 2366-3413 (electronic)
Sustainable Civil Infrastructures
ISBN 978-3-030-79671-6 ISBN 978-3-030-79672-3 (eBook)
https://doi.org/10.1007/978-3-030-79672-3

This Springer imprint is published by the registered company Springer Nature Switzerland AG
The registered company address is: Gewerbestrasse 11, 6330 Cham, Switzerland

Introduction

This volume contains eight papers that were accepted and presented at the GeoChina 2021 International Conference on Civil and Transportation Infrastructures: From Engineering to Smart and Green Life Cycle Solutions in Nanchang, China, September 18–19, 2021. It features novel tunneling technologies and analytical and numerical methods for behavior analysis and modeling of soils and underground infrastructure. A number of important topics are covered, including methodological and pragmatic solutions to critical issues in tunnel engineering (e.g., soil arching and invert heaving), investigations on soil behavior (e.g., penetration resistance of mono-bucket foundations in silty soil and inception of debris avalanches), as well as novel approaches to infrastructure condition survey and inspection based on point cloud data and infrared image analysis. It is anticipated that this collective information will lead to improved safety and resilience in design and construction, and efficiency and sustainability in maintenance and management of underground infrastructure.

Contents

About the Editors

Prof. J. James Yang is an associate professor of Civil Engineering in the College of Engineering at University of Georgia. He received his Ph.D. degree from the University of South Florida. He is a professionally registered transportation engineer with a research interest in sustainable and resilient infrastructure systems, smart mobility systems, statistical and econometric methods, artificial intelligence and machine learning methods and applications. His work has generated several patents and over 40 publications in refereed journals and conference proceedings. He currently sits on the Editorial Board of ASCE Journal of Infrastructure Systems, the TRB AED50 committee: Artificial Intelligence and Advanced Computing Applications, and the Transportation Demand Management Coordinating Committee (TDM-CC) of Atlanta Regional Commission.

Professor Wen-Chieh Cheng is a professor of Geotechnical Engineering in the School of Civil Engineering at Xi'an University of Architecture and Technology. He received his Ph.D. degree from the National Taipei University of Technology. He is a professionally registered civil engineer with a research interest in sustainable and resilient infrastructure systems, soil contamination and remediation strategies, trenchless methods, artificial intelligence and machine learning methods and applications. His work has generated several international patents and over 70 publications in refereed journals and conference proceedings. He currently sits on the Editorial Board of ICE Geotechnical Research journal, and the ISSMGE TC213 committee: Scour and Erosion.

Dr. Shuying Wang is a professor of Tunnel Engineering in School of Civil Engineering, Central South University. He obtained his B.E. and M.S. degrees from Central South University, respectively, in 2005 and 2007 and then earned the Ph.D. degree in Missouri University of Science and Technology (Rolla, USA) in 2011. His research interests include conditioning mechanism and handling technology of special soils, tunneling mechanism and safety control technology. He has been hosting about 30 research projects, one of which is National Natural Science Foundation-Outstanding Youth Foundation. So far, more than 70 papers have been published, and 14 patents have been authorized. Additionally, he has gotten one monograph and one textbook pressed in Springer. He won the second prize of science and technology progress of Ministry of Education in 2019 (in the first place) and that of Hunan Province in 2017 (in the third place). He serves as an editorial board member for three journals such as Journal of Testing and Evaluation (ASTM), Chinese Journal of Highway, etc.

Inception of Debris Avalanches: A Material Point Method Modelling

Sabatino Cuomo[1] (✉), Angela Di Perna[1], and Mario Martinelli[2,3]

[1] University of Salerno, Salerno, Italy
scuomo@unisa.it
[2] Deltares, Delft, Netherlands
[3] TU Delft, Delft, Netherlands

Abstract. Rainfall-induced landslides of the flow type in granular soils are among the most complex natural hazards due to the variety of mechanisms which regulate the failure and propagation stages. Among these, debris avalanches are characterised by distinct mechanisms which control the lateral spreading and the increase in soil volume involved during the propagation. Two different stages can be individuated for debris avalanches, i.e. the failure stage and the avalanche formation stage: the former includes all the triggering mechanisms which cause the soil to fail; the latter is associated to the increase of the unstable volume. Regarding these issues, in the literature, either field evidence or qualitative interpretations can be found while few experimental laboratory tests and rare examples of geomechanical modelling are available for technical and/or scientific purposes.

In this paper a contribution is provided about the advanced numerical modelling of the inception of such hazardous debris avalanches. Particularly, the case of the impact of a failed soil mass on stable deposits is considered. This means that a small translational slide occurs; the failed mass causes the soil liquefaction of further material by impact loading; the landslide volume increases inside triangular-shaped areas during the so-called "avalanche formation", and also soil erosion along the landslide propagation path plays an important role.

To this aim, an innovative numerical technique known as the Material Point Method (MPM) is used. It can be considered as a modification of the well-known Finite Element Method (FEM) particularly suited for large deformations. The continuum body is schematized by a set of Lagrangian points, called Material Points (MPs). Large deformations are modelled by MPs moving through a background mesh, which also covers the domain where the material is expected to move. The MPs carry all physical properties of the continuum such as stress, strain, density, momentum, material parameters and other state parameters, whereas the background mesh is used to solve the governing equations without storing any permanent information. Such advanced approach allows combining a hydro-mechanical coupled approach, any of the well-known soil constitutive models proposed over the years in soil mechanics and a large-displacement formulation.

The numerical analyses are performed adopting 2D geometrical configurations taken from field evidences and previous researches. Triangular 3-noded computational meshes are used, characterized by elements of about 1 m. The interaction between the impacting mass, and then of the propagating flow with the in-situ stable soil is examined, providing important insights about the behaviour

J. J. Yang et al. (Eds.): GeoChina 2021, SUCI, pp. 1–15, 2021.
https://doi.org/10.1007/978-3-030-79672-3_1

of such type of landslide. The results achieved so far are encouraging and show that MPM can properly simulate the inception of debris avalanches and even their complex mechanisms during the impact and the interaction with in-situ stable zones.

1 Introduction

Flow-type landslides are among the most complex natural hazards due to their high velocities and long run-out distances and also to the variety of mechanisms that occur within failure and post-failure stages (Pastor et al. 2009; Cascini et al. 2012; Cuomo 2020). In the specific, the so-called "debris avalanches" still pose significant open issues in the scientific debate, since a consistent mathematical framework for their analysis has not yet been formulated.

According to the most recent landslide classification (Hungr et al. 2014), the term "debris avalanche" states for "very rapid to extremely rapid shallow flow of partially or fully saturated debris on a steep slope, without confinement in an established channel". They typically occur in open slopes, i.e. shallow soil deposits with almost constant depths and slope angles generally between 30° and 45°. In particular, debris avalanches start with small involved volumes (failure stage) and then turn into larger landslides because of the increasing in mobilized volume through further failures or eventual soil entrainment and resulting as a triangular-shaped area (post-failure stage).

Additional to that, soil unsaturated condition may be a key factor, in the failure and post-failure stages. Before failure, the additional strength related to matric suction is fundamental for the equilibrium of granular soil slopes steeper that the effective friction angle. During the failure stage of rainfall-induced landslides, the soil suction gradually reduces due to rain infiltration and peculiar mechanical responses can even occur such as the capillary collapse (strong reduction of soil volume related to wetting). But also during the propagation stage soil suction may evolve towards higher or smaller values depending on the amount of deformation in the landslide body. Thus, it would be desirable to have a comprehensive hydro-mechanical-coupled and large-displacement-based approach to include and accurately analyse all of these issues.

Referring to failure and post-failure stages, four different zones can be distinguished (Fig. 1). Zone 1 corresponds to small failures that occur at natural or anthropogenic discontinuities of soil deposits (respectively, bedrock outcrops and cut slopes). Zone 2 is the impact zone of the previously mentioned failed masses that usually corresponds to water supplies from bedrock (either karst spring or water runoff at bedrock outcrops); if zone 1 is absent, zone 2 is the source area of small landslides triggered by water supplies from bedrock. Zone 3 corresponds to distinct mechanisms: thrust of the failed mass upon the downslope stable material and/or soil entrainment due to the propagating mass. Zone 4 exclusively corresponds to soil entrainment. It is worth noting that while zones 1 and 2 are few tens of metres large, the width of zones 3 and 4 is not known a priori and its forecasting is a challenging task.

Generally, for analysing the triggering of a landslide, classical approaches like Limit Equilibrium Methods (LEMs) are often employed, which completely neglect the soil deformations and rely only on equilibrium equations under simplified hypotheses. Alternatively, or in conjunction to that, stress-strain analyses through Finite Element Methods

(FEMs) have been also performed, considering the soil deformations generally "small". This simplification may be a reasonable hypothesis when the pre-failure and the failure are the only issues of the analysis. In addition, the hydro-mechanical coupling between the solid skeleton and the pore water pressure can be rigorously taken into account.

On the other hand, the propagation stage of such kind of landslides has been mostly analysed in terms of soil displacements, but not so much in terms of hydromechanical coupling during soil evolution. The simulation of the propagation stage during slope instability was managed through several approaches, such as discrete element method (Cuomo et al. 2019; Zhao et al. 2019) or Lagrangian particle-based methods such as SPH, PFEM, FEMLIP, MPM (Cuomo et al. 2013, 2015; Ceccato et al. 2018a; Cuomo 2020; Yuan et al. 2020).

In this paper, an effort is done to provide a conceptual and computational tool useful to perform different types of analysis in order to capture the essential aspects of complex hydro-mechanical behaviour during the inception and evolution of a debris avalanche following the impact of an unstable mass. To this aim, MPM proves to be a powerful method, able in simulating the complex mechanics of landslide motion during the failure, propagation and deposition stages.

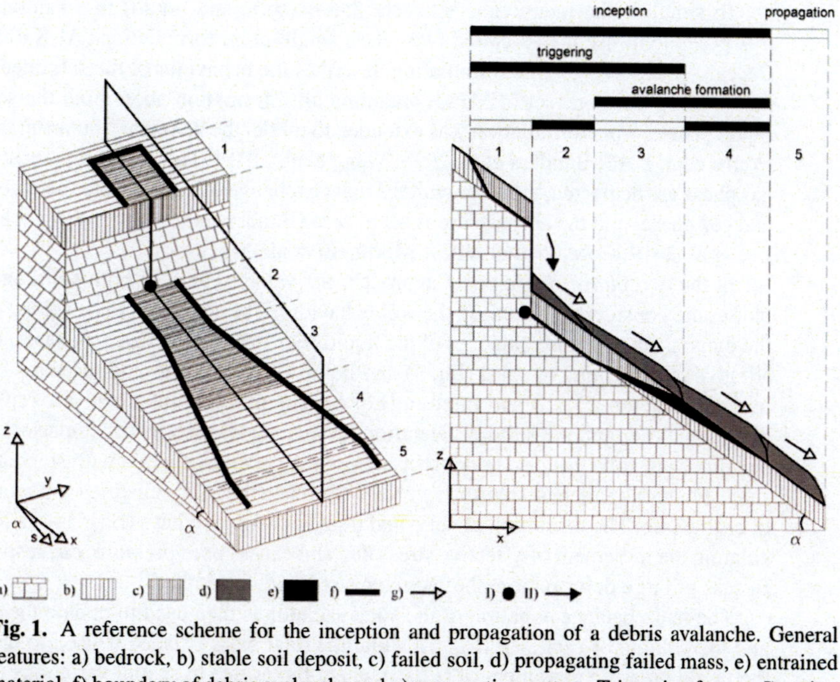

Fig. 1. A reference scheme for the inception and propagation of a debris avalanche. General features: a) bedrock, b) stable soil deposit, c) failed soil, d) propagating failed mass, e) entrained material, f) boundary of debris avalanche and g) propagation pattern. Triggering factors: I) spring from bedrock, II) impact loading. *Zone 1–2* triggering; *zone 3* thrust of failed material and/or soil entrainment; *zone 4* soil entrainment, zone 5 propagation (Cascini et al. 2012).

Considering the relevance of the topic, essentially related to the destructiveness of these phenomena, an advanced modelling of debris avalanches could provide a valuable tool to improve (i) the understanding of the inception (i.e. triggering mechanism and avalanche formation) of debris avalanches, (ii) the capability to forecast these landslides and (iii) the evaluation of the mobilised volume inside the avalanche source area, which is a crucial point for hazard and risk assessment and zoning.

2 Materials and Methods

2.1 The Material Point Method (MPM)

The Material Point Method (MPM) can be considered as an enhancement of the finite element method (FEM) since it is well suited for large deformation problems (Sulsky et al. 1994). The continuum body consists of a number of Lagrangian points (named material points, MPs), which carry all the physical properties and stress-strain values. The MPs movement through a background fixed mesh which allows modelling large deformations of the body. The mesh is used to solve the governing equations without storing any permanent information.

To simulate the interaction between phases (solid and liquid) in a saturated soil, it is possible to use the so-called two-phase single-point formulation (Al-Kafaji 2013; Ceccato et al. 2018b). This formulation describes the behaviour of the saturated porous media using only one set of MPs, containing all information about both the solid and liquid phases. This formulation was extended to model the behavior of unsaturated soils (Yerro et al. 2015; Bandara et al. 2016; Wang et al. 2018). The governing equations for gas phase are neglected, together with the mass exchange of air and water between liquid and gas phase, and the air pressure is set to zero (Bandara et al. 2016; Wang et al. 2018; Lee et al. 2019; Ceccato et al. 2021; Martinelli et al. 2021).

In the *two-phase single-point* approach, the velocity field of both solid and liquid phases are considered to analyse the soil behaviour. The equations to be solved concern the dynamic momentum balances of the liquid (Eq. 1) and solid (Eq. 2) phases, the constitutive relationships of solid (Eq. 3) and liquid (Eq. 4) phases and the mass balances of solid skeleton (Eq. 5) and mixture (Eq. 6). The primary unknowns are represented by the liquid $\boldsymbol{a}_l$ and the solid acceleration ($\boldsymbol{a}_l$ and $\boldsymbol{a}_s$, respectively). Numerically, these equations are solved at grid nodes considering the Galerkin method with standard nodal shape functions and their solutions are used to update the MPs velocities and momentum of each phase. The resolution of solid and liquid constitutive laws (Eqs. 3–4) allows calculating the increment of effective stress $\mathrm{d}\boldsymbol{\sigma}'$ and excess pore pressure $\mathrm{d}\boldsymbol{u}$, respectively. In case of large deformation, the Jaumann stress rate is adopted.

The mass balance equation of the solid skeleton is then used to update the porosity and the volume of each MP (Eq. 5), while the total mass balance is used to determine the liquid pressure increment of (Eq. 6) where fluxes due to spatial variations of liquid mass are neglected ($\nabla n\rho_l = 0$).

All the equations are based on the assumption that solid grains are incompressible. Furthermore, the flow is considered laminar and stationary for slow velocity regime. This means that Darcy's law can be applied to describe the interaction between solid and liquid phases (i.e. drag force $\boldsymbol{f}_d$ in Eq. 7).

$$\rho_l \boldsymbol{a}_l = \nabla u - \frac{n\mu_l}{k}(\boldsymbol{v}_l - \boldsymbol{v}_s) + u\boldsymbol{b} \tag{1}$$

$$(1-n)\rho_s \boldsymbol{a}_s = \nabla \cdot (\boldsymbol{\sigma} - nu\boldsymbol{I}) + (\rho_m - n\rho_l)\boldsymbol{b} + \frac{n\mu_l}{k} n(\boldsymbol{v}_l - \boldsymbol{v}_s) \tag{2}$$

$$\mathrm{d}\boldsymbol{\sigma}' = \mathbf{D} \cdot \mathrm{d}\boldsymbol{\varepsilon} \tag{3}$$

$$\mathrm{d}\boldsymbol{u} = \mathbf{K}_l \cdot \mathrm{d}\boldsymbol{\varepsilon}_{vol,l} \tag{4}$$

$$\frac{Dn}{Dt} = (1-\mathrm{n})\nabla \cdot \boldsymbol{v}_s \tag{5}$$

$$\frac{D\boldsymbol{\varepsilon}_{vol,l}}{Dt} = \frac{(1-\mathrm{n})}{n}\nabla \cdot \boldsymbol{v}_s + \nabla \cdot \boldsymbol{v}_l \tag{6}$$

$$\boldsymbol{f}_d = \frac{n\mu_l}{k}(\boldsymbol{v}_l - \boldsymbol{v}_s) \tag{7}$$

In the Eqs. (1–7), ρ_l is the liquid density, u is the liquid pressure, μ_l is the liquid dynamic viscosity, k is the intrinsic permeability of the liquid, v_s and v_l are the solid and liquid velocity, $\boldsymbol{b}$ is the body force vector, n is the soil porosity, ρ_s is the solid density, ρ_m is the density of the mixture ($\rho_m = \rho_s\ (1-n) +\ n\rho_l$), $\boldsymbol{\sigma}$ is the total stress tensor of the mixture, $\mathbf{D}$ is the tangent stiffness matrix, $\mathbf{K}_l$ is the elastic bulk modulus of the liquid phase and $\boldsymbol{\varepsilon}_{vol,l}$ is the volumetric strain.

The simplified two-phase single-point formulation for unsaturated soils extends the above-mentioned approach considering the effect of the degree of saturation S_l, so the dynamic momentum balances of the liquid phase and the mixture are those reported in Eqs. (8) and (9), respectively. In this paper, the formulation illustrated in Lee et al. (2019) and Martinelli et al. (2021) is used. The reader can refer to the original paper for further details.

The term k_{rel} is the ratio between the soil permeability related to suction through a characteristic curve and the saturated permeability, hence it is equal to 1 for a fully saturated soil and it decreases with the degree of saturation. However, in this study k_{rel} is set constant to 1. Furthermore, the Bishop relationship to compute the effective stress is considered to include the effect of the partially saturation in the soil behaviour (10).

Considering the assumption of a barotropic liquid phase, the time derivative of the liquid density is only function of time derivative of the pore water pressure (11). Since gas properties are neglected, the suction (s) is equal to $-u$.

The water retention curve is obtained through a linearised relation for simplicity (12), in which a_v is a constant value.

$$\rho_l \boldsymbol{a}_l = \nabla u + \frac{nS_l\mu_l}{k_{rel}k_l}(\boldsymbol{v}_l - \boldsymbol{v}_s) + u\boldsymbol{b} \tag{8}$$

$$(1-n)\rho_s \boldsymbol{a}_s = \nabla \cdot (\boldsymbol{\sigma} - nS_l u\boldsymbol{I}) + (\rho_m - nS_l\rho_l)b + \frac{nS_l\mu_l}{k_{rel}k} nS_l(\boldsymbol{v}_l - \boldsymbol{v}_s) \tag{9}$$

$$\boldsymbol{\sigma}' = \boldsymbol{\sigma} - S_l u \boldsymbol{I} \quad (10)$$

$$\frac{\partial u}{\partial t} = \frac{K_l}{\left(1 - \frac{K_l}{S_l}\frac{\partial S_l}{\partial s}\right)}\left[\frac{(1-n)}{n}\nabla \cdot \boldsymbol{v}_s + \nabla \cdot \boldsymbol{v}_l\right] \quad (11)$$

$$S_l = 1 - a_v s \quad (12)$$

2.2 Scheme and Input Data

Some simplified calculation schemes were considered to analyse the different mechanisms leading to the inception of a debris avalanche from the impact of a mobilized mass of a slope above. The general reference scheme is represented in Fig. 2, in which the unstable mass (Material 1) is in a fully saturated condition while the soil on the slope downstream (Material 2) is stable in unsaturated condition due to the presence of soil suction (s_0). The unstable mass with uniform depth $H_{slope,1}$, length $B_{slope,1}$ and slope i_1 falls from a height H and crashes on the stable slope with depth $H_{slope,2}$, length $B_{slope,2}$ and overall inclination i_2.

Some relevant cases were considered to understand the influence of different geometries and boundary conditions on the triggering and propagation stages, as reported in Table 1. In particular, the role of suction and the length extension of the Material 2 have been investigated in this study.

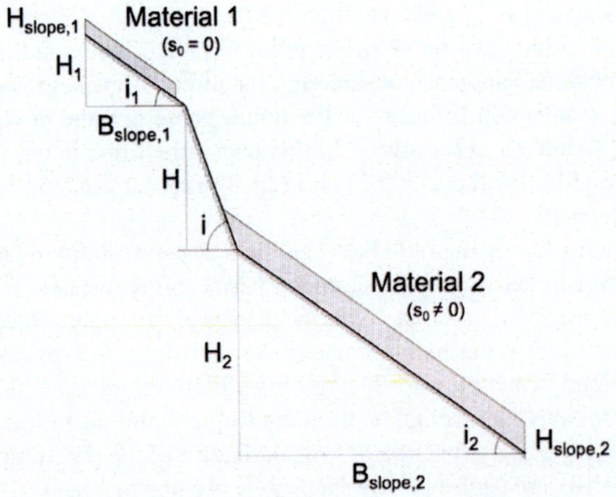

Fig. 2. Reference scheme for the numerical analyses.

Suction value and soil properties, such as shear strength, stiffness, saturated permeability and unit weight (Table 2), are those typical of the coarser superficial ashy soils (silty sands), having been widely investigated by Bilotta et al. (2005). On the other hand, the saturated hydraulic conductivity and water bulk modulus are typical of fine sands.

With regard to the material constitutive models, both Materials 1 and 2 were modelled using a linear elastic-perfectly plastic behaviour. Furthermore, the water retention characteristic curve for the unsaturated material was set as linear function for simplicity, with parameters $a_v = 0.01$, whereas relative permeability curve is neglected and so hydraulic conductivity was kept constant.

The boundary condition $u = 0$ is set along the ground level, leading to a decrease in suction over long time due to consolidation. In any case, the analyses last a very short period (some tens of seconds), so the consolidation phase has no influence.

Table 1. Simulation cases and related geometric/boundary features.

	H_1(m)	$H_{1,slope}$(m)	$B_{1,slope}$(m)	i_1(°)	H (m)	I (°)	H_2(m)	$H_{2,slope}$ (m)	$B_{2,slope}$ (m)	i_2(°)	s_0(kPa)
Case 1	10	1.50	14	35	40	70	73	3	105	35	5
Case 2	10	1.50	14	35	40	70	34.5	3	45	35	5
Case 3	10	1.50	14	35	40	70	73	3	105	35	0

Table 2. Material properties used in the numerical analyses.

	ρ_{sat}(kg/m^3)	N (–)	K_0(–)	E (MPa)	v(–)	c′ (kPa)	φ′ (°)	φ′$_b$(°)	ψ′(°)	ksat (m/s)	K_L (MPa)	μ_L (Pas)	a_v(–)
Material 1	1315	0.58	0.38	5	0.29	0.1	38	38	0	10^{-3}	50	10^{-3}	-
Material 2	1315	0.58	0.38	5	0.29	5.0	38	38	0	10^{-3}	50	10^{-3}	0.01

3 Modelling the Impact-Induced Debris Avalanche

The numerical analyses were performed adopting a 2D geometrical configuration taken from field evidences and previous researches (Cascini et al. 2012; Cuomo et al. 2013). Triangular 3-noded elements computational meshes are used, characterized by elements with average size of 1 m (Fig. 3). The contact between soils and base material is managed through a frictional contact algorithm (Al-Kafaji 2013). The boundary conditions are set for both liquid and solid phase. The velocities are fully fixed in both directions along all the boundaries of the domain, whereas the right edge of the Material 2 is fixed only in the X direction.

The initial stress state of the soils was initialized assuming a geostatic condition, with a k0 coefficient set to 0.38. The water table was set at the bottom for the partially saturated material with a constant suction profile at the beginning of the simulation.

The numerical results aimed to assess the time-space evolution of some quantities, such as stress, strain, pore pressure and velocities, for all the above-mentioned cases. Firstly, the spatial distribution for some relevant time lapses was considered. The results show that the impact of the mass in the Case 1 (Fig. 4) apparently has no effect in the first moments (up to about 50 s), as well shown by the velocity distribution in the material downstream, which remains in stable condition. However, the degree of saturation is increasing during this period and a perched water table is forming in the lower part of the slope. After that, the material becomes unstable due to the increasing in pore-water pressure and reaches high velocities (over 6 m/s).

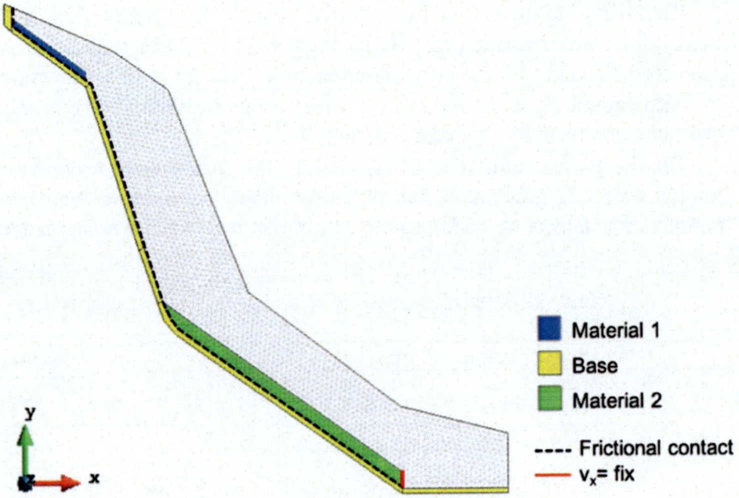

Fig. 3. Computational domain and materials.

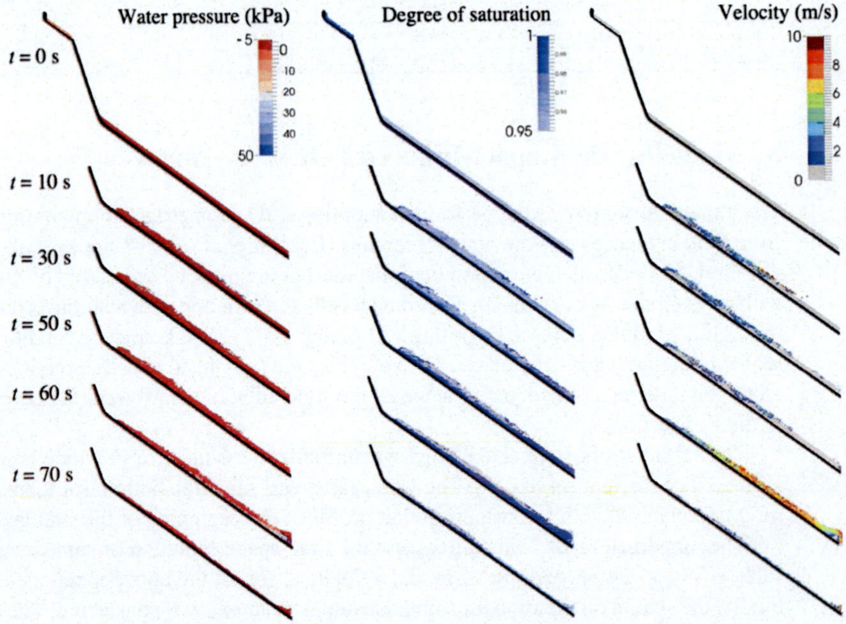

Fig. 4. Spatial distribution of water pressure, degree of saturation and velocity (Case 1).

Similar results were achieved for the Case 2 (Fig. 5), even if the reduced length of the stable material causes a delayed failure compared to Case 1. In fact, soil begins to move at about 85 s against the 60 s of the previous case. Furthermore, the velocity reached by the moving mass does not exceed 4 m/s.

Completely different is the case in which suction is not taken into account (Fig. 6). The impacting mass suddenly leads the below material to instability, due to the increase of pore-water pressure at impact. The material behaves like a flow which quickly propagates downstream, reaching very high velocity (over 10 m/s) during the simulation. It is worth noting that in this case there is no delay for the formation of the avalanche, which develops in few seconds.

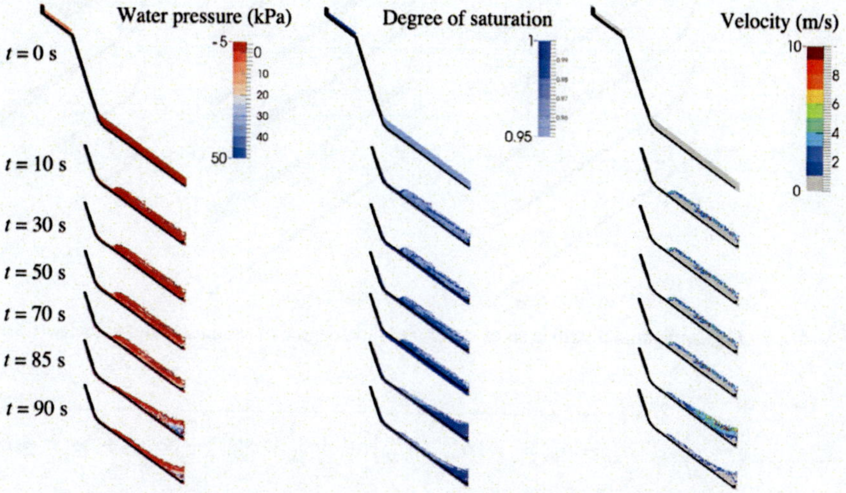

Fig. 5. Spatial distribution of water pressure, degree of saturation and velocity (Case 2).

Additional to the analysis of the spatial distribution of some relevant quantities, it may be useful to investigate their trend over time in specific zones of the impacted mass. The elements of the mesh taken as reference fixed zones are 5, equally spaced and at half depth. The numerical results are displayed only if at least one material point is present in the element. If the element becomes empty (e.g. the material points leave the element after failure) no temporal trend is shown in the plots.

In order to analyse the stress state condition, the mean effective stress (p'), the total mean stress (p), deviatoric stress (q) and water pressure (u) were considered, as defined in Eqs. (13–15). In particular, the ratio q/p' was compared to the critical state (Eq. 16). Both q/p' and u/p were considered over time since they are dimensionless quantities and so more easily comparable for the three reference cases. In conjunction, the velocity pattern and the deviatoric strain ε_q (Eq. 17) were also considered to add more information about the inception and propagation mechanisms.

$$p' = \frac{\sigma'_{xx} + \sigma'_{yy} + \sigma'_{zz}}{3} \tag{13}$$

$$p = p' + u \tag{14}$$

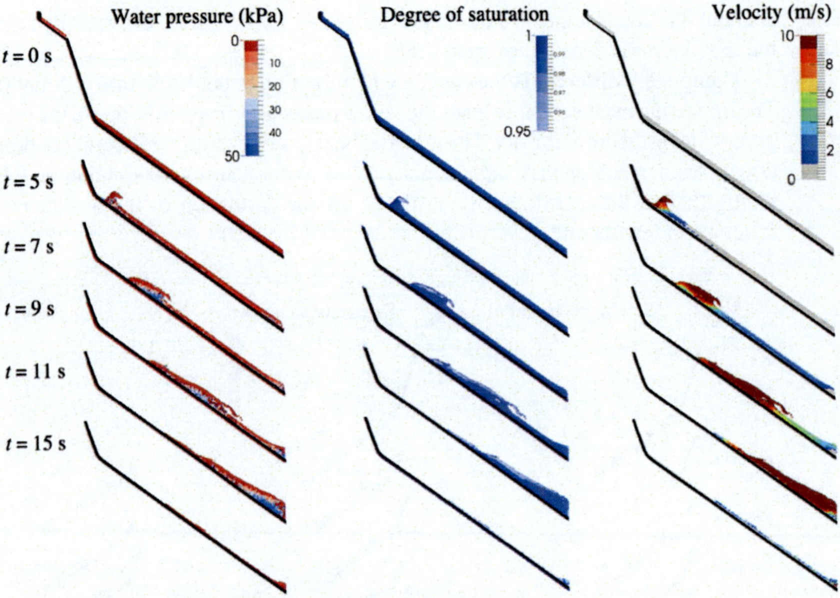

Fig. 6. Spatial distribution of water pressure, degree of saturation and velocity (Case 3).

$$q = \frac{1}{\sqrt{2}}\sqrt{\left(\sigma'_{xx} - \sigma'_{yy}\right)^2 + \left(\sigma'_{yy} - \sigma'_{zz}\right)^2 + \left(\sigma'_{xx} - \sigma'_{zz}\right)^2 + 6{\sigma'_{xy}}^2 + 6{\sigma'_{yz}}^2 + 6{\sigma'_{zx}}^2} \tag{15}$$

$$M = \frac{6sin\varphi'}{3 - sin\varphi'} \tag{16}$$

$$\varepsilon_q = \frac{\sqrt{2}}{3}\sqrt{\left(\varepsilon_{xx} - \varepsilon_{yy}\right)^2 + \left(\varepsilon_{yy} - \varepsilon_{zz}\right)^2 + \left(\varepsilon_{xx} - \varepsilon_{zz}\right)^2 + 6{\varepsilon_{xy}}^2 + 6{\varepsilon_{yz}}^2 + 6{\varepsilon_{zx}}^2} \tag{17}$$

Comparing these above-mentioned ratios for Case 1 (Fig. 7), it emerges that following the impact of the unstable mass, the stable material reaches the critical line for an instant (excluded point E, which doesn't seem to notice the impact) then the slope quickly finds a new equilibrium configuration. In the meantime, pore-water pressure is increasing, starting from the lower zone (point E). Suddenly, the q/p' ratio reaches the critical value M in all points and so the slope begins to move, as also shown by the trend of velocity and deviatoric strain (Fig. 8), which increases from below upward.

Even for Case 2 (Fig. 7), the stable material reaches the critical state for a while but in this case also in point E, due to the reduced length of the slope. The only exception is point B which still remains in yielding condition, but without affecting the equilibrium of the whole mass. After reaching the new equilibrium configuration in the other zones, also here pore-water pressure increases, but this time more markedly from the above (point B). After few tens of seconds, the q/p' ratio goes beyond the critical line M in the central zone (points B, C and D), where the highest velocity and deviatoric strain are reached (Fig. 8).

Completely different is the Case 3 (Fig. 7), in which q/p' is very close to M just from the beginning, due to the absence of suction. Then, following the impact, the mass begins quickly to move from the above (point B) and propagates downward very fast, as shown by the velocity trend (Fig. 8). While the mass is propagating, the points that are farther away from the triggering become more stable due to the compression wave, until their q/p' ratio crosses the critical value. Differently from the other cases, in Case 3 the shear strain evolve over time with the propagation of the unstable mass, as well shown in Fig. 8.

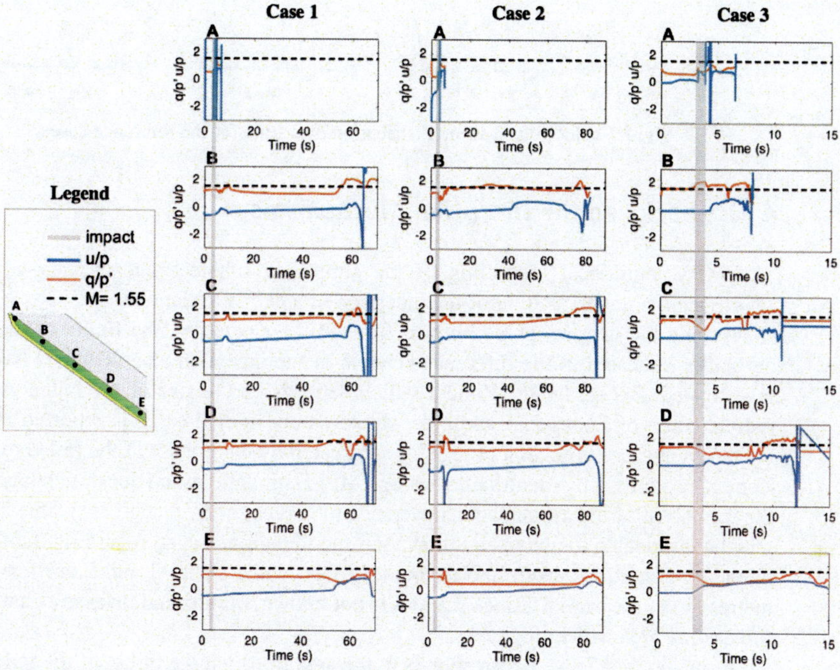

Fig. 7. Stress distribution over time for the reference cases.

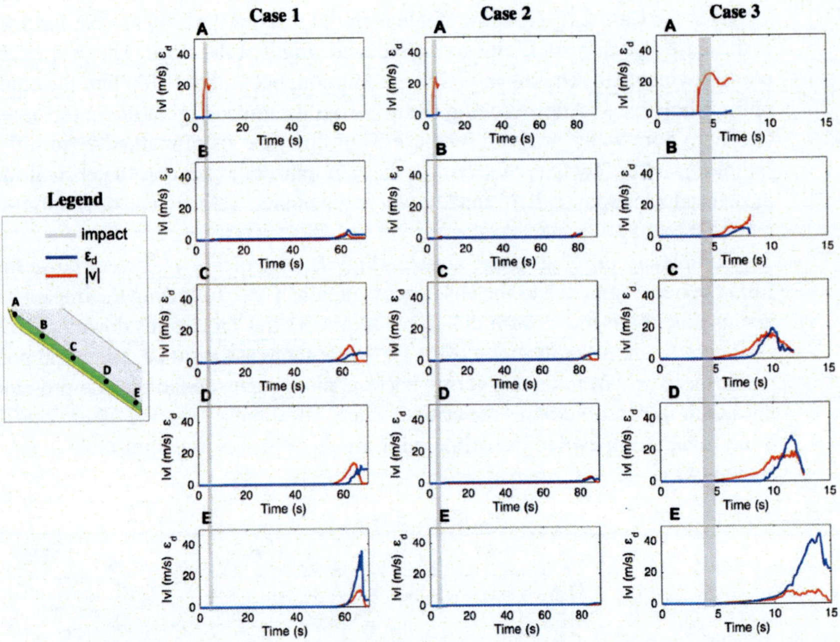

Fig. 8. Velocity and strain distribution over time for the reference cases.

4 Remarks on the Inception Mechanisms

The MPM numerical modelling of the simplified schemes above-analysed allows improving the understanding of the inception of a debris avalanche.

Among the new insights obtained from MPM modelling, the limits of the different zones indicated in the reference scheme of the inception stage (Fig. 1) have been investigated. Except for the Zone 1 which depends on the size of the fallen mass, the individuation of Zones 2, 3 and 4 for the three considered cases is reported in Fig. 9 for different cases. The area of the slope triggered by the impact of the failed mass (i.e. Zone 2) is quite easily identifiable through MPM simulations and doesn't change during the formation of the avalanche. Conversely, the other zones change over time and so it may be reasonable to understand how their modification occurs, regarding the different boundary conditions taken into account. In fact, while Zones 1 and 2 are few tens of metres large, the size of Zones 3 and 4 is not known a priori and its assessment and/or forecasting are challenging tasks.

Knowing that Zone 3 corresponds to the area in which the thrust of the failed mass and/or soil entrainment occur and Zone 4 is exclusively related to soil entrainment, their individuation in some crucial instants can be made. In fact, what emerges from Fig. 9 is that for all the cases the length of Zone 3 ranges from few meters (when the most part of the downslope material is stable) to many tens in the final configuration. The increase of the initial failed volumes can be associated with further failures occurring inside this zone. The transition from initial slide to a real debris avalanche is also related to soil entrainment along the path of the failed mass (Zone 4). In particular, the comparison between the reference cases in Fig. 9 shows that this zone is very elongated if suction is present, while it is shorter but involves more failed volumes in the Case 3, where suction is not taken into account. Relevant remarks can also be made by comparing Cases 1 and 2 in terms of time. For instance, at the same instant of 70 s, the deposited material along the slope in Case 1 is representative of Zone 3 only, while Case 2 is characterized by both Zones 3 and 4, highlighting a sort of delay in the formation of these zones.

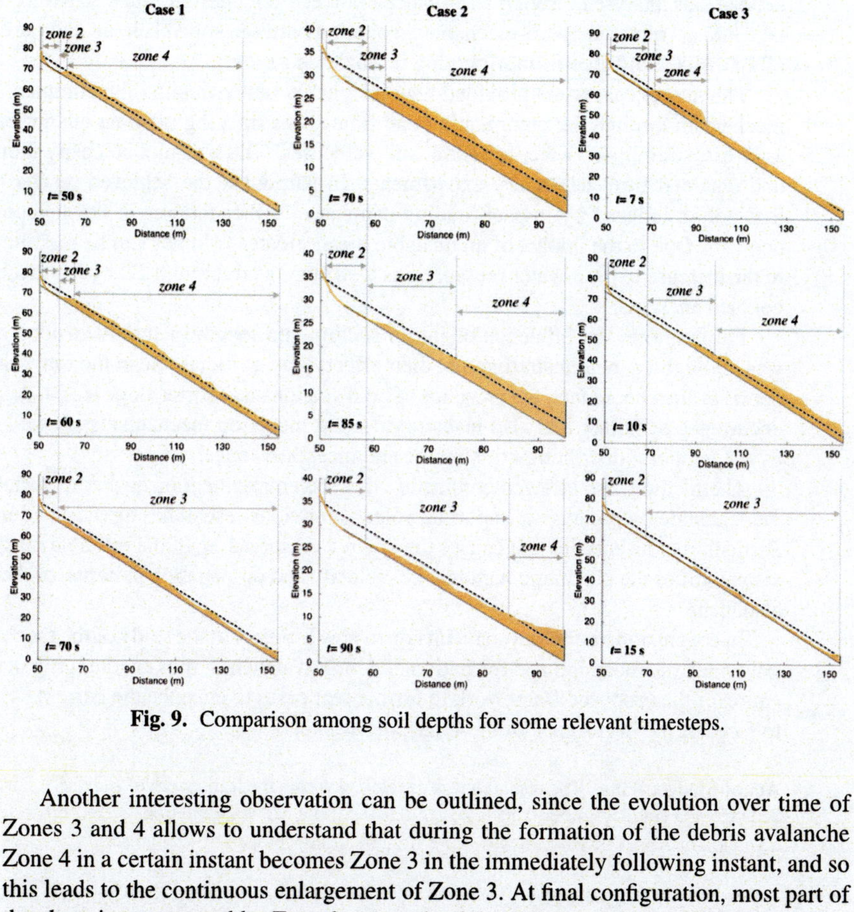

Fig. 9. Comparison among soil depths for some relevant timesteps.

Another interesting observation can be outlined, since the evolution over time of Zones 3 and 4 allows to understand that during the formation of the debris avalanche Zone 4 in a certain instant becomes Zone 3 in the immediately following instant, and so this leads to the continuous enlargement of Zone 3. At final configuration, most part of the slope is represented by Zone 3, as result of the soil entrainment, while most failed volumes included in Zone 4 during propagation have turned into a stationary deposit beyond the edge of the slope.

In conclusion, the entrainment of further material makes the propagation patterns complex, so the use of an advanced numerical technique like MPM is useful to provide new insights in the understanding of the debris avalanches inception mechanisms.

5 Concluding Remarks

Debris avalanches are complex natural hazard for which several field evidences or qualitative analyses are provided in the current literature, even if few examples of laboratory tests and/or geomechanical modelling are available for this kind of phenomena.

The study proposed herein aimed to provide a general overview of the current potentialities for the geomechanical modelling of such dangerous phenomena. In particular, the inception of a debris avalanche due the impact of a failed soil mass on a stable deposit was analysed through an advanced numerical modelling, such as MPM, able to take into account the hydro-mechanical coupling between solid skeleton and pore-water of a flowlike landslide from triggering to final deposition.

The numerical results provided new insights in understanding the complex impact mechanism through the computation and time-space tracking of some quantities, such as stress, strain, pore-water pressure, soil velocities, which cannot be easily monitored and obtained from laboratory experiments. In particular, the achieved results outline that failure induced by impact loading can cause further failures in downslope stable deposits. Due to the impact of an unstable mass, greater volumes can be mobilized due to the increase in pore-water pressures, as well simulated through 2D hydro-mechanical coupled analyses.

Furthermore, the influence of initial suction and length of the downslope deposit was highlighted, demonstrating that their effects play a crucial role in the formation of a debris avalanche. In fact, the presence of both suction and shorter slope length delays the occurrence of failure and also leads to different inception mechanisms, as well shown by the temporal distribution of stresses and inception zones.

The method also allowed confirming that the entrainment of further material along the landslide path plays an important role, as previous researches already had noticed. Particularly, the entrainment causes the downward spreading of the unstable mass and it also modifies the kinematic features, i.e. velocity and propagation patterns, of the debris avalanche.

In conclusion, the achieved results open new frontiers to the study of the mechanisms governing the inception and formation of debris avalanches, thus encouraging the application of the proposed framework to further real cases to enhance the current capability to forecast the occurrence of these hazardous phenomena.

Acknowledgements. The research was developed within the framework of Industrial Partnership PhD Course (POR Campania FSE 2014/2020). All the MPM simulations were performed using a version of Anura3D (http://www.mpm-dredge.eu/) developed by Deltares.

References

Al-Kafaji, I.K.A.: Formulation of a dynamic material point method (MPM) for geomechanical problems. PhD. Thesis. University of Stuttgard (2013)

Bandara, S., Ferrari, A., Laloui, L.: Modelling landslides in unsaturated slopes subjected to rainfall infiltration using material point method. Int. J. Numer. Anal. Meth. Geomech. **40**(9), 1358–1380 (2016)

Bilotta, E., Cascini, L., Foresta, V., Sorbinow, G.: Geotechnical characterisation of pyroclastic soils involved in huge flowslides. Geotech. Geol. Eng. **23**(4), 365–402 (2005)

Cascini, L., Cuomo, S., Pastor, M.: Inception of debris avalanches: remarks on geomechanical modelling. Landslides **10**(6), 701–711 (2012). https://doi.org/10.1007/s10346-012-0366-0

Ceccato, F., Redaelli, I., di Prisco, C., Simonini, P.: Impact forces of granular flows on rigid structures: comparison between discontinuous (DEM) and continuous (MPM) numerical approaches. Comput. Geotech. **103**, 201–217 (2018)

Ceccato, F., Yerro, A., Martinelli, M.: Modelling soil-water interaction with the material point method. evaluation of single-point and double-point formulation. Num. Methods Geotech. Eng. **IX**, 351–357(2018b)

Ceccato, F., Yerro, A., Girardi, V., Simonini, P.: Two-phase dynamic MPM formulation for unsaturated soil. Comp. Geotech. 129 (2021)

Cuomo, S.: Modelling of flowslides and debris avalanches in natural and engineered slopes: a review. Geoenviron. Disast. **7**(1), 1–25 (2020). https://doi.org/10.1186/s40677-019-0133-9

Cuomo, S., Calvello, M., Villari, V.: Inverse analysis for rheology calibration in SPH analysis of landslide run-out. In: Lollino, G., Giordan, D., Crosta, G.B., Corominas, J., Azzam, R., Wasowski, J., Sciarra, N. (eds.) Engineering Geology for Society and Territory – Volume 2, pp. 1635–1639. Springer, Cham (2015). https://doi.org/10.1007/978-3-319-09057-3_291

Cuomo, S., Moretti, S., Aversa, S.: Effects of artificial barriers on the propagation of debris avalanches. Landslides **16**(6), 1077–1087 (2019). https://doi.org/10.1007/s10346-019-01155-1

Cuomo, S., Prime, N., Iannone, A., Dufour, F., Cascini, L., Darve, F.: Large deformation FEMLIP drained analysis of a vertical cut. Acta Geotech. **8**(2), 125–136 (2013)

Fern, J., Rohe, A., Soga, K., Alonso, E. (eds.): The Material Point Method for Geotechnical Engineering: A Practical Guide. CRC Press, Boca Raton (2019)

Hungr, O., Leroueil, S., Picarelli, L.: The Varnes classification of landslide types, an update. Landslides **11**(2), 167–194 (2013). https://doi.org/10.1007/s10346-013-0436-y

Lee, W., Martinelli, M., Shieh, C.-L.: Modelling rainfall-induced landslides with the material point method: the Fei Tsui Road case. In: Proceedings of the XVII ECSMGE-(2019)

Martinelli, M., Rohe, A.: Modelling fluidisation and sedimentation using material point method. In: 1st Pan-American Congress on Computational Mechanics (2015)

Martinelli, M., Lee, W.L., Shieh, C.L., Cuomo, S.: Rainfall Boundary Condition in a Multiphase Material Point Method. In: Tiwari, B., Sassa, K., Bobrowsky, P.T., Takara, K. (Eds.) Understanding and Reducing Landslide Disaster Risk. WLF 2020. ICL Contribution to Landslide Disaster Risk Reduction, pp. 303–309. Springer, Cham (2021). https://doi.org/10.1007/978-3-030-60706-7_29

Pastor, M., Haddad, B., Sorbino, G., Cuomo, S., Drempetic, V.: A depth-integrated, coupled SPH model for flow-like landslides and related phenomena. Int. J. Numer. Anal. Methods Geomech. **33**(2), 143–172 (2009)

Sulsky, D., Chen, Z., Schreyer, H.L.: A particle method for history-dependent materials. Comput. Methods Appl. Mech. Eng. **118**(1–2), 179–196 (1994)

Yerro, A., Alonso, E., Pinyol, N.M.: The material point method for unsaturated soils. Géotechnique **65**(3), 201–217 (2015)

Yuan, W.-H., Liu, K., Zhang, W., Dai, B., Wang, Y.: Dynamic modeling of large deformation slope failure using smoothed particle finite element method. Landslides **17**(7), 1591–1603 (2020). https://doi.org/10.1007/s10346-020-01375-w

Zhao, L., Liu, X., Mao, J., Shao, L., Li, T.: Three-dimensional distance potential discrete element method for the numerical simulation of landslides. Landslides **17**(2), 361–377 (2019). https://doi.org/10.1007/s10346-019-01282-9

Wang, B., Vardon, P.J., Hicks, M.A.: Rainfall-induced slope collapse with coupled material point method. Eng. Geol. **239**, 1–12 (2018)

Research on the Calculation of Segment Floating Considering the Action Degree of Soil Arching Effect

Daxin Geng[1,2], Yuchen Hu[1(✉)], Yalong Jiang[1,2], and Ning Wang[1,2]

[1] School of Civil Engineering and Architecture, East China Jiaotong University, Nanchang 330013, Jiangxi, China
gdaxin@ecjtu.edu.cn

[2] National Experimental Teaching Demonstration Center of Civil Engineering, East China Jiaotong University, Nanchang 330013, China

Abstract. Based on the background of the mud shield tunnel engineering in the section between Anfeng Station and Dongxin Station of Line 4 of Nanchang metro, the stress composition of the segment during the shield construction process was analyzed firstly, which was composed of buoyancy, the weight of the segment and the slurry, the viscous resistance of the segment, the friction resistance between the rings, the shear resistance of the bolt and the force of the overlying soil. Secondly, establishing the corresponding mechanical model, wherein: when calculating the viscosity resistance caused by the slurry, the slurry was assumed to be Newtonian; when calculating the overburden pressure, the degree of soil arching effect was considered, and the modified Terzaghi loosening earth pressure calculation was used. Through theoretical analysis of the stress of the segment during the construction phase, the theoretical solution of the amount of the segment from the shield tail to the initial settling of the slurry could be deduced, and then compared the theoretical solution and the basic field monitoring data. Finally, effective anti-buoyancy measures were proposed for the floating of the segments. The result shows that the theoretical solution of the floating amount derived by the calculation model is closer to the actual value, which verifies the feasibility of the calculation model and proposes targeted anti-floating measures based on the force composition of the segment, so as to achieve a good control effect of the upward moving of segments.

Keywords: Shield tunnel · Segment floating · Soil arching effect · Loosening earth pressure of Terzaghi · Calculation model · Anti-floating measures

1 Introduction

With the rapid development of China's economic and the increasing pressure of urban traffic, the construction of subway is becoming more and more important to meet the demand of urban traffic. Shield method is widely used in subway tunnel construction

D. Geng—Mainly engaged in basic engineering, tunnel engineering and other aspects of research.

J. J. Yang et al. (Eds.): GeoChina 2021, SUCI, pp. 16–36, 2021.
https://doi.org/10.1007/978-3-030-79672-3_2

because of its obvious advantages such as small ground interference, high construction efficiency and environmental friendliness [1]. In the process of shield tunneling, when the segment comes out of the shield tail, a circular shield tail gap will be formed between the segment and the surrounding soil mass. If synchronous grouting and secondary filling are not carried out in time, the segment will float locally or as a whole, further resulting in the segment dislocation, cracking and even serious it will deviate from the designed excavation axis [2, 3].

Nowadays, many academics at home and abroad have made a lot of research on the calculation and analysis of the stress of segments, also they have made lots of achievements. Taking the comprehensive pipe corridor of Caofeidian Industrial Zone as the background, Based on the bending differential equation and bending moment coordination equation of elastic foundation beam, Wang Daoyuan et al. deduced the theoretical solution of longitudinal uplift of shield tunnel under water; Based on the project of "Shenxianshu station to South Railway Station" of Chengdu Metro Line 7, Zhang Jun et al. [6] have classified the upward moving of segment into four categories and proposed the calculation method of segment floatation from the end of shield to the initial setting of slurry; Based on Terzaghi's loosening earth pressure theory, Li Chunlin [7] deduced a calculation formula of loosening earth pressure which can consider the formation loss and segment stiffness; Based on the Shiziyang project in the Pearl River, Lu Qianqian et al. [8] considered the effect of foundation resistance and derived the calculation model of segment buoyancy through the theory of force balance; Taking the shield tunnel project of a certain variable composite stratum in Nanning subway as the background, Shuyao et al. [9] used the longitudinal equivalent continuous beam model to predict and analyze the upward moving of segment among different typical stratum; based on the consideration of the constraint of adjacent segments on the floating segments and the compression effect of residual force on the overlying soil, Ye Fei et al. [10] calculated and analyzed the anti-floating problem of shield tunnels; Based on the shield tunnel project of the west line of Shanghai North cross passage, Yuan Wei et al. [11] analyzed the stress of segment structure in the construction process and established the calculation model of segment buoyancy after the initial setting of slurry; Zhang Jun et al. [12] analyzed the stress state of segment structure in the construction process, and proposed the floating calculation method from the end of segment out of shield to the initial setting of slurry. The main difference in the above analysis in deriving the calculation formula of the upward moving of segment is the calculation of the force of the overlying soil layer on the segment. At present, there are two main calculation methods: 1) Vertical load calculation method. That is to say, the overlying soil layer only bears the buoyancy resistance of the segment within the width of the tunnel, and the self-weight load of the overlying soil directly acts on the segment; 2) Calculation of Terzaghi loosening earth pressure. It is based on Terzaghi's collapse arch theory and from the point of stress transfer, the vertical earth pressure is calculated by considering the soil arch effect in the process of tunnel excavation.

The excavation of the shield causes the soft soil to relatively slip and form a shearing effect, thereby it makes the cohesive force productive. In the vicinity of the excavation surface, due to the large space loss of the stratum, a hollow area is generated above the shield, which will cause the soil to undergo active deformation, thus providing conditions for the occurrence of soil arching. The soft soil stratum mainly transmits the load generated by the shield excavation through the cohesive force between the particles and the friction force. Under the external force of active excavation, the cohesive force between the soft soil particles drops sharply, then soil arching forms in the deep stratum. The formation of soil arching effect mainly depends on the soil displacement. Therefore, thc stratum loss caused by deep stratum excavation determines the initial displacement of soil arching conditions. Due to the large burial depth of the shield, the soil settlement caused by the stratum loss rate will not extend to the surface indefinitely.

The ratio corresponding to the soil arching effect is the ratio of the earth pressure generated by the active excavation at the arching crown of the tunnel to the initial earth pressure. For the cross-river shield tunnel. The core of the subway tunnel load calculation is mainly the calculation of the vertical taxi pressure at the top of the tunnel, the soil arching effect and load of the overlying stratum of the tunnel are important which is associated with the design and construction of shield tunnels. Xu Weizhong et al. studied the influence of shield depth on soil arching effect and found that the bigger the buried depth of the shield tunnel in soft soil is, the more effective the soil arch effect is. During the unloading process of deep soft soil stratum, the displacement of the soil makes the soil particle density of the soil arch boundary increase. But the relationship between the proportion of soil arching effect and the depth of tunnel is non-linear. The soil arching effect plays an important role when the excavation of the shield in the depth of 2 times diameter or deeper area in soft soil. The soil arching effect can reach 23% when the depth reaches 4D. The research focuses on relationship between the depth of embedment and the proportion of soft soil arching effect, which has a certain significance value in vertical load for soft soil tunnel. Therefore, the buried depth of shield tunnel has a certain influence on the degree of soil arching effect.

When calculating the upward moving of the segment, the soil arching effect is an important factor in calculating the force of the overlying soil layer on the segment. Considering the soil arching effect, the theoretical calculation of the loosening earth pressure of the Terzaghi foundation is generally adopted at present. However, the degree of soil arching effect is not considered, so the traditional Terzaghi loosening earth pressure formula is not very adaptable in shield tunnels, it is mainly based on the tunnel construction method of mining.

This paper relies on the river crossing tunnel project from Anfeng station to Dongxin station of Nanchang rail transit line 4, considering the effect of the degree of soil arch effecting and the weight of the slurry on the buoyancy resistance of the segment, the purpose is to adopt the modified Terzaghi foundation which can consider the degree of soil arching effect, the use the loosening earth pressure formula to calculate the force of the overlying soil stratum on the segment, we can establish a model to calculate the upward moving of the segment and carry out efficient anti-floating measures which were based on the calculation results on site, furthermore it can suppress the floating of the lining segment to a certain extent.

Based on the specific engineering background, the theoretical solution of segment buoyancy can be obtained through the analysis of segment floating. On the one hand, some anti-floating measures can be taken to control the segment floating as effectively as possible from the perspective of mechanical analysis. On the other hand, the theoretical solution of segment floating can be predicted before segment erection, and the erection position can be adjusted properly during segment erection after the segment detach from shield tail, the segment erection quality will not be affected due to the large difference in the upward moving of adjacent segments, or even lead to segment rupture. Therefore, according to the theoretical value of segment buoyancy, the distance between the segment erection axis and the design axis can be adjusted reasonably. In order to ensure that the tunnel axis is on the design axis after the segment comes out of the shield tail, so as to ensure the segment erection quality.

2 Analysis of Segment Floating

During the process of shield construction, the upward moving of the segment is caused by multiple factors, such as: buoyancy of segment, construction gap, construction disturbance and driving parameters [4]. This section analyzes the buoyancy of the segment.

2.1 The Effect of Groundwater

When the shield tunnels in water rich stratum or saturated soil with good water permeability, the whole tunnel will be surrounded by water, and the segment will be subjected to a greater water buoyancy after it is released from the shield tail. If the buoyancy of water is greater than the soil load and the self-weight of segment, then the segment will float upward.

2.2 The Impact of the Tunnel Wall Grouting and Grouting Slurry

2.2.1 The Impact of the Tunnel Wall Grouting

The slurry diffusion after grouting behind the wall is a very complex process, which is related to the surrounding formation parameters, slurry liquid ratio, grouting pressure and grouting construction, in fact, the main reason for it is grouting construction. The most direct impact of slurry diffusion on the segment is to make the segment subject to the "dynamic buoyancy".

The slurry diffusion can be divided into four stages, namely filling stage, infiltration stage, compaction stage and splitting stage. The compaction stage refers to pressing the high concentration slurry into the soil layer with high pressure. At this time, a "slurry bubble" forms near the bottom of the grouting pipe and the "slurry bubble" is an important factor for the "dynamic buoyancy" of the segment [13].

2.2.2 The Impact of Grouting Slurry

When the shield machine is driving forward, the excavation diameter of the shield is larger than the outer diameter of the segment, so the segment will form a circular gap with the surrounding soil layer after it is pulled out from the shield tail, which is called the shield tail gap. In order to maintain the stability of the tunnel during the shield construction stage, it is necessary to carry out synchronous grouting for shield tail gap, then the slurry will wrap the segment and generate buoyancy.

2.3 The Impact of Back Channeling of Mud Slurry

In the process of slurry shield construction, it is necessary to apply large slurry pressure to balance the pressure of water diagram at the incision, so as to ensure the stability of the excavating face. When slurry shield tunneling in hard rock stratum, the surrounding rock converges and wraps the lining segment at a slower speed. It is easy to form seepage and slurry channel between surrounding rock and lining, which is likely to cause slurry back channeling and increase buoyancy.

2.4 The Impact of Rebound Action of Foundation

The cutter-head will cut the soil in front of the shield when the shield machine is driving, which will lead to the unloading of the foundation and the rebound of the foundation. At the same time, the segment will be subjected to a vertical foundation reaction. Generally speaking, the buoyancy of the segment is mainly composed of the resilience of the foundation, water buoyancy, grouting construction behind the wall and slurry buoyancy [10].

3 Stress Analysis of Segment

3.1 Establishment of Calculation Model

During the procedure of shield excavation, since the excavation diameter of the shield is greater than the outer diameter of the segment, a circular building gap will be formed between segment and surrounding soil after segment is separated from the shield tail. In order to maintain the stability of the tunnel during construction, synchronous grouting needs to be implemented in this gap, and the main reason for the rise of the segment is that the large buoyancy would generate during the grouting procedure. In the period of segment buoyancy, the segment buoyancy is greater than its buoyancy resistance, in the meanwhile, the segment buoyancy increases gradually. but as the slurry gradually condenses, the buoyancy of the segment decreases gradually, and finally the buoyancy is equal to the anti-buoyancy that it receives, then the segment will reach a static equilibrium state. At the time, the upward moving of the segment reaches the maximum value, which is the final floating amount of the segment. The stress diagram of shield segment is shown in Fig. 1.

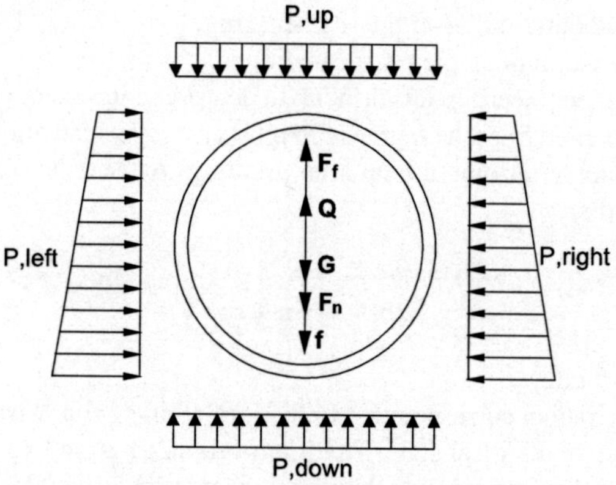

Fig. 1. Stress diagram of segment

When the segment floats to the final stability, the static balance equation of segment can be obtained according to the buoyancy resistance equal to the buoyancy.

$$F_{\mathrm{f}} + P,\ down = G + F_{\mathrm{n}} + f + P, up \text{ - } Q_{\mathrm{max}} \tag{1}$$

Where:

F_f—Buoyancy on the segment;
P, down—Foundation reaction force on segment;
G—Dead weight of segment and slurry;
F_n—Viscous resistance of segment;
F—friction resistance between segments;
P, up—The force of the overlying soil layer on the segment.

3.2 Calculation of Buoyancy

Ye Fei pointed out that the buoyancy is mainly composed of the "static buoyancy" generated by the slurry and water wrapped segment and the grouting construction disturbance [13].

3.2.1 Calculation of the "Static Buoyancy"

When the segment subjects to "static buoyancy", the calculation expression of "static buoyancy" can be obtained according to Archimedes buoyancy calculation formula, the "static buoyancy" calculation expression is:

$$F_{\mathrm{f1}} = \gamma_L V_{\mathrm{c}} = \gamma_L \pi R_1^2 b \tag{2}$$

Where:
F_{f1}—Static buoyancy of segment;
γ_L—Floatation of the liquid around the segment;

R_1—The outer radius of the segment ring;

b—The width of single ring segment.

In actual engineering, it is difficult to determine the exact value of the volume of the slurry discharged from the segment. In most cases, you can only know the corresponding arc of the slurry forming a loop after grouting. After improving formula (2), it can be concluded that:

$$\begin{aligned} F_{f1} &= \gamma_L V_p \\ &= \gamma_L R_1^2 b\left(\tfrac{\alpha}{2} - \sin\tfrac{\alpha}{2}\cos\tfrac{\alpha}{2}\right) \end{aligned} \quad (0 \le \alpha \le 2\pi) \tag{3}$$

among them:

α—The radian corresponding to the formation of slurry ring. It can be seen that the radian is 2π in the ideal state. The improved calculation expression is consistent with the formula obtained from Archimedes' calculation principle.

3.2.2 Calculation of the "Dynamic Buoyancy"

In the process of construction, it is inevitable that the segments are staggered and damaged, which will affect the stress of segments to a certain extent. It is assumed that the segments are idealized during the stress analysis of the segments. Because Nanchang Metro Line 4 is located in water-rich sand layer zone, only the effects of the infiltration stage and the compaction stage are considered when the calculation is simplified. In actual projects, most of the segments float up locally, which is mainly caused by the local concentration of grouting slurry. Therefore, the local distribution mode is used in the calculation and analysis. Based on the adverse principle, when the grouting slurry gathers at the lower part of the segment, a large concentration force or distribution force can be formed [14]. The dynamic floating calculation diagram of segment is shown in Fig. 2.

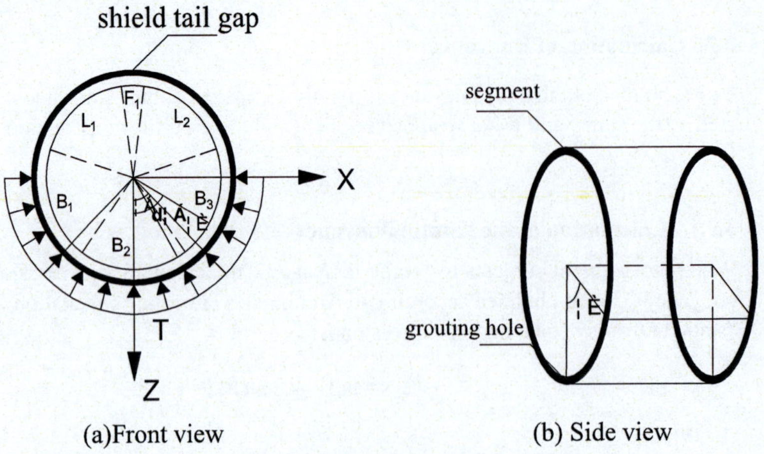

Fig. 2. Calculation model of "Dynamic Buoyancy" of segment

"Dynamic buoyancy" is the component force of grouting pressure in the vertical direction, which is calculated as follows:

$$\begin{aligned} F_{f2} &= 2\int_0^{\theta} TA\cos\alpha d\,\alpha\ \theta(0<\theta<\frac{\pi}{2}) \\ &= 2\int_0^{\theta} TR_1 b\cos\alpha\ d\alpha = 2TR_1\mathrm{b}\sin\theta \end{aligned} \tag{4}$$

Where:

F_{f2}— "Dynamic buoyancy" of segment;

T—Grouting pressure;

θ—The angle between the far end of slurry diffusion and the vertical direction. Based on the most unfavorable factors, $\theta = \pi/2$ is taken, then $F_{f2} = 2TR_1b$.

3.2.3 Calculation of Foundation Reaction

The cutterhead will cut the soil in front of the shield machine when the shield machine is driving forward, which can lead to the unloading of the foundation and the rebound of the foundation. At the same time, the segments will be subject to a vertical reaction of the foundation. However, Because the influence of formation loss and the influence caused by foundation rebound appear to be mutual elimination, the reaction of foundation may not be considered in the stress analysis of segment [15], that is, assuming P, down = 0.

3.2.4 Establishment of the Buoyancy Equation

The slurry will gradually harden after synchronous grouting behind the segment wall. As the slurry coagulates, the buoyancy of segment decreases gradually. Li Yunli carries out model experiment on segment buoyancy of shield tunnel, and finds that the buoyancy resistance of segment decreases linearly with time [16], that is to say, the buoyancy equation is as follows:

$$\begin{aligned} F_f &= F_{f1} + F_{f2} - kt \\ &= \gamma_L \pi R_1^2 b + 2TR_1\mathrm{b} - kt \end{aligned} \tag{5}$$

In the formula:

K—Reduction in buoyancy per unit time, which can be taken as 19.67 kN/h based on experience [13].

3.3 Calculation of Anti-buoyancy

3.3.1 Calculation of the Dead Weight of Segment and Slurry

$$G = \sum_{i=1}^{2} \gamma v = \gamma_c \pi (R_1^2 - R_2^2) b + \gamma_L \pi (R^2 - R_1^2) b \tag{6}$$

In the formula:

γ_c—Unit Weight of reinforced concrete;

R_2—Inner radius of segment ring;

g—Gravitational acceleration.

R—Excavation radius of shield.

3.3.2 Viscous Resistance of Segment

Because of the formation of a slurry flow layer around the segment after grouting, the slurry flow layer will cause the segment to suffer some viscous resistance. Assuming that the slurry is Newtonian, the Newtonian internal friction law is satisfied, that is to say, the shear stress of the fluid is directly proportional to the shear deformation rate and the angular deformation rate [11], which satisfies the formula as follows:

$$\tau = \mu \frac{du}{dy} = \mu \frac{d\theta}{dt} \left(N/m^2, pa\right) \tag{7}$$

Where:
τ—Shear stress;
μ—Dynamic viscosity;
u—Velocity of flow.

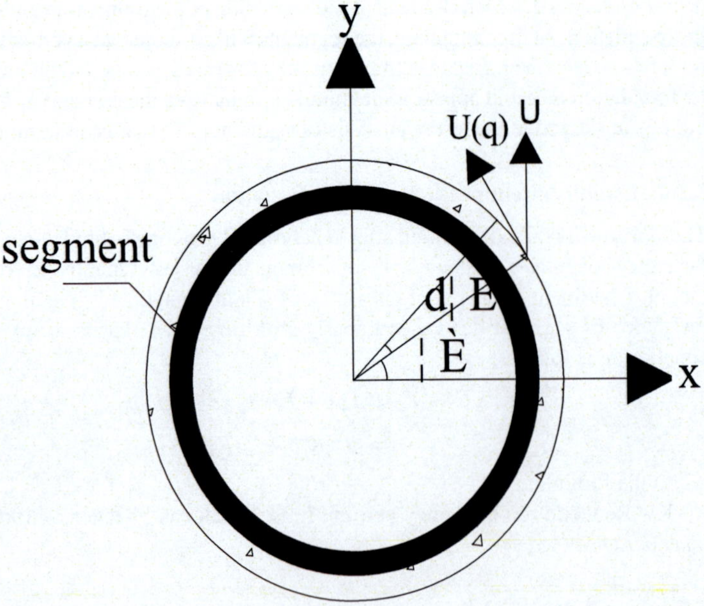

Fig. 3. Calculation model of viscous resistance

According to the Newtonian plate test, the shield tail gap is regarded as the Newtonian plate model, the velocity between plates is linearly distributed along the Y direction, and the calculation model of the viscous resistance of segments is shown in Fig. 3.

The expression of slurry shear stress is:

$$\tau = \mu_t \frac{du}{dy} = \mu_t \frac{U(q)}{\delta} \tag{8}$$

Where:
μ_t—Coefficient of slurry viscosity;
U(q)—The tangential velocity of the slurry;
δ—Thickness of shield tail gap.
When the grouting is completed, there is a relationship between the tangential velocity of the slurry and the floating velocity of the segment:

$$U(y) = U \cdot \cos\theta \tag{9}$$

In the formula:
U—Floating speed of the segment;
θ—The tangential velocity and vertical angle of the slurry.
The expression of the viscous resistance of the segment:

$$\begin{aligned} F_n &= 2\int_0^{\pi} \tau \text{ A} \cos\theta d\theta = 2\int_0^{\pi} \tau \text{ R}_1 \, b \cos\theta d\theta \\ &= 2\int_0^{\frac{\pi}{2}} \tau \, \mu_t \frac{\frac{dS}{dt}}{\delta} \int_0^{\pi} \text{R}_1 \, b \cos\theta d\theta \\ &= 4\frac{\mu_1 R_1 b}{\delta}\frac{dS}{dt}\frac{1}{2}\frac{\pi}{2} = \frac{\mu_1 R_1 b}{\delta}\frac{dS}{dt} \end{aligned} \tag{10}$$

Considering the viscosity coefficient of slurry as Newtonian time-varying function, namely:

$$\mu = k_1 e^{k_2 t} \tag{11}$$

Where:
k_1 and k_2 are the time-varying parameters of cement mortar, which are seen as 2.5 kPa·h and 0.02 respectively. Then we can get the viscous resistance of the segment:

$$F_n = \frac{\pi \text{k}_1 e^{k_2 t} R_1 b}{\delta}\frac{dS}{dt} \tag{12}$$

3.3.3 Friction Resistance of the Segment

Considering the size deviation between the bolt and the bolt hole wall, the corresponding anti-floating calculation should be analyzed in two stages, namely, the ring friction stage and the bolt shear stage. The friction between segments can be calculated according to the following formula:

$$f = \mu\left(\sum_{i=1}^{n} N_i + N_j\right) \tag{13}$$

In the formula:

μ—The friction coefficient between the rings, the approximate value is 0.3;
N_i—Pre-tightening force of single bolt;
N_j—The residual pressure from the jacking thrust force to the ring is 2000 kN based on experience.

3.3.4 Calculation of Bolt Shear Strength

The segment thickness of shield tunnel is far less than the outer radius of segment ring, which accounts for about one-twentieth. Therefore, the segment ring section can be regarded as a thin-walled ring section in the shear analysis of bolts. The maximum shear stress occurs on the neutral axis, its direction is collinear with the shear direction, and the value can be calculated according to formula (14):

$$\tau_{\max} = \frac{QS'_{x\max}}{I_x b'} \tag{14}$$

In the formula:
Q—Shear force of segment ring;
S,xmax—The moment of inertia of half a segment ring on the neutral axis;
I_x—The moment of inertia of the segment ring to the neutral axis;
t—The thickness of segment;
b´—The width of the segment ring at the neutral axis, b´ is equal to 2t;

$$I_x = \frac{\pi}{4}[(R_a + \frac{t}{2})^4 - (R_a - \frac{t}{2})^4] \approx \pi t R_a^3 \tag{15}$$

$$S'_{x\max} = 2tR_a^2 \tag{16}$$

In the formula:
R_a—The average radius of the segment ring.
Substituting formula (14), we can get:

$$\tau_{\max} = \frac{Q}{\pi t R_a} \tag{17}$$

And then we can get the maximum shear stress formula on the segment:

$$Q_{\max} = \frac{\tau_{\max}}{2} N\pi r_b^2 = [\tau]\frac{N\pi r_b^2}{2} \tag{18}$$

In the formula:
[τ]—Allowable shear stress of bolts;
N—The number of longitudinal bolts;

3.3.5 The Action of Overlying Soil on the Segment

Terzaghi proved the existence of the soil arch effect through the Trapdoor test in 1936. Based on the limit equilibrium theory of loosening media, Terzaghi proposed a calculation formula for loosening earth pressure, that is, the soil will only produce a shear failure surface when it is below a certain critical depth [17]. The stress of surrounding rock is often released during the tunnel excavation, which will cause uneven displacement or relative displacement between the soil, and then the soil arching effect in the soil. In Terzaghi theory, it is assumed that the soil is in the limit equilibrium state. In actual engineering, it is often necessary to constrain the displacement of the soil. For example, in the shield tunnel, the soil arch effect cannot fully function because the shear stress of the soil is less than its shear strength. The traditional Terzaghi loosening earth pressure formula is not very adaptable in shield tunnels. The traditional Terzaghi loosening earth pressure theory is mainly based on tunnels constructed by the mining method, but it is not very applicable in the shield method [7]. The expression of Terzaghi loosening earth pressure is:

$$P, up = \sigma_V = \frac{B\left(\gamma - \frac{c}{B_t}\right)}{K\tan\varphi}\left(1 - e^{-K\tan\varphi\frac{h}{B_t}}\right) + P_0 e^{-K\tan\varphi\frac{h}{B_t}} \tag{19}$$

In the formula:
c—Cohesion;
B_t—Half of the width of the loosening area;
K—Coefficient of lateral pressure;
γ—Average weight of overlying soil;
h—The thickness of overlying soil;
φ—Angle of internal friction.
Among them:

$$B_t = R_1 \cot\left(\frac{\pi}{8} + \frac{\varphi}{4}\right) \tag{20}$$

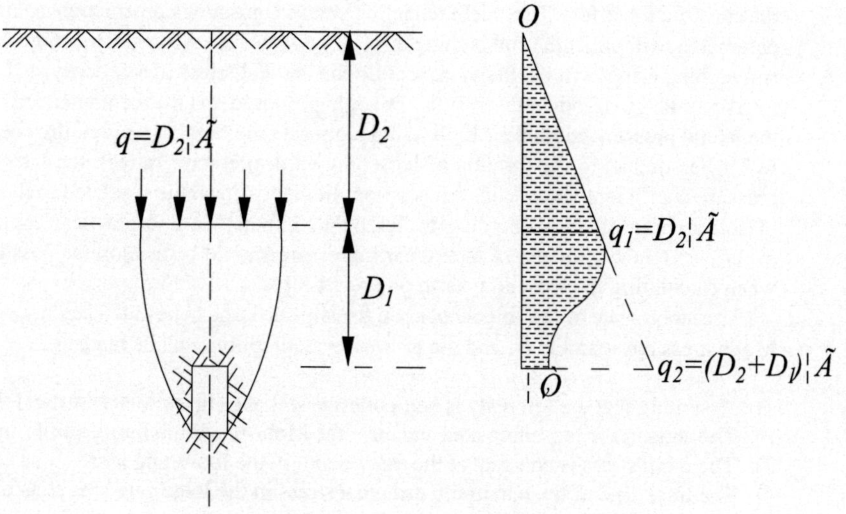

Fig. 4. Stress contour plots of soil in loosening area

When the buried depth of the tunnel is large, the soil arching effect can not be extended along the surface infinitely, the soil arching effect only exists in the vault D_1. Kezdi believes that D_1 value is generally taken as 1.5–3 times of diameter for circular tunnel, the part above the dome D1 is regarded as an equal settlement surface, that is, the shear stress in the soil is 0, there is no soil arching effect, and it is regarded as a distributed load. The loosening earth pressure profile is shown in Fig. 4, based on this, the formula of Terzaghi's loosening earth pressure should be written in the following formula: [19].

$$\sigma_{\text{v}} = \frac{B_t\gamma - c'}{K\tan\varphi}\left(1 - e^{-K\tan\varphi\frac{D_1}{B_t}}\right) + (P_0 + \gamma D_2)\, e^{-K\tan\varphi\frac{D_1}{B_t}} \tag{21}$$

In the formula:

c′—Cohesion between soil particles that actually plays a role due to the relative movement tendency. When the soil arch effect is fully effective, c′ is equal to c, when the soil arch effect does not exist, then c′ is equal to 0, It can be seen that the value of c′ is related to the degree of soil arch effect, and the value is between 0 and c [18].

Although formula (21) considers soil arching effect, it does not consider the degree of soil arching effect. Based on this, Li Chunlin deduced the formula of loosening earth pressure under incomplete soil arching effect, the formula can fully reflect the action degree of soil arching effect:

$$\sigma_{\text{v}} = \frac{B_t\gamma - c'}{A_1}\left(1 - e^{-A_1\frac{D_1}{B_t}}\right) + (_0 + \gamma D_2)e^{-A_1\frac{D_1}{B_t}} \tag{22}$$

Through comparison, it can be seen that when $A_1 = k{\cdot}tan\varphi$, it is the formula of Terzaghi loosening earth pressure. Li Chunlin pointed out that $k{\cdot}tan\varphi$ is not a constant in practice. The value depends on the degree of soil arching effect, and is related to the construction effect. The shield tunneling stratum, grouting construction and material parameters will affect the soil arching effect. The larger the A1 value, the more fully the soil arching effect will be, the more rapidly the vertical stress in soil decays [7].

According to trapdoor test results, Terzaghi proposed that the recommended value of the lateral pressure coefficient K_1 is 1. Terzaghi set the lateral earth pressure coefficient to 1 when deducing the formula of loosening earth pressure. In fact, the lateral earth pressure coefficient varies with the position. In theory, its value is related to the degree of soil arching effect. According to Terzaghi's experience and practical engineering experience, the coefficient of lateral earth pressure should be as close as possible to 1 when calculating the loosening earth pressure [20].

The stress state of the soil element on the slip surface can be calculated by using the Mohr stress circle in Fig. 5, and the following assumptions can be made:

1) Assuming that the soil body is non-cohesive soil, the slip surface is vertical through;
2) The soils in the loosening area can meet the Mohr-Coulomb limit equilibrium state;
3) The main stress is constant at the same depth in the loosening area;
4) The track line of the maximum principal stress in the loosening area is an arch line with the opening downward.

From the geometric relationship in Fig. 5, it can be concluded that:

$$\delta_{\text{v1}} = \delta_1 \sin^2\theta + \delta_3 \cos^2\theta \tag{23}$$

$$\delta_{\text{h}} = \delta_1 \cos^2\theta + \delta_3 \sin^2\theta \tag{24}$$

Dividing Eq. (19) by Eq. (18):

$$K = \frac{\delta_h}{\delta_v} = \frac{1 + K_a \tan^2 \theta}{K_a + \tan^2 \theta} \tag{25}$$

Where k is the coefficient of lateral pressure at any unit point in the soil layer in the loosening area; $K_a = \delta_3/\delta_1$ is the coefficient of active earth pressure, it follows that $Ka = \tan^2(\pi/4\text{-}\varphi/2)$, it can be seen from the formula that when $\theta = \pi/2$, $K = K_a$; Then the passive earth pressure coefficient is introduced, when $\theta = 0, K_p = \tan^2 (\pi/4 + \varphi/2)$, we can find that $K = 1/K_a = K_p$, the value of the lateral pressure coefficient K at the unit point is exactly equal to k_p, it can be seen that the side pressure coefficient K at any unit point floats between the active earth pressure coefficient Ka and the passive earth pressure coefficient K_p, that is to say, the soil pressure coefficient is related to the value of θ, The existence of the earth arching effect makes the coefficient of the lateral pressure of the soil be a variable value. The higher the degree of soil arching effect, the greater the side pressure coefficient, and the failure surface is $\theta = \pi/4 + \varphi/2$. In view of this, Krynine proposed the side pressure coefficient $K' = (1-\sin2\varphi)/(1 + \sin2\varphi)$.

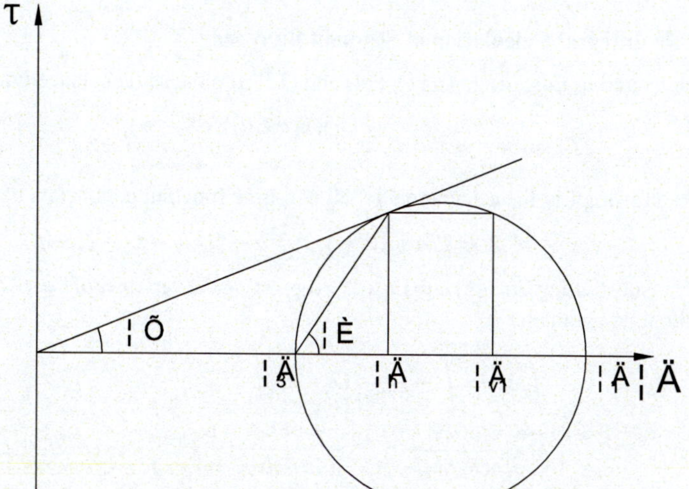

Fig. 5. Stress state of soil in loosening area

Draw the relationship diagram of various lateral pressure coefficients under different friction angles in the soil, as shown in Fig. 6.

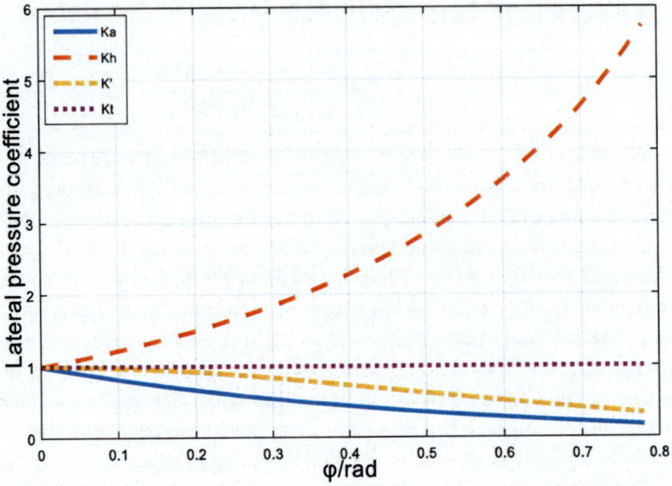

Fig. 6. Diagram of relationship between various lateral pressure coefficients and friction angle in soil

3.4 Integral Calculation of Segment Buoyancy

Substituting Eqs. (4), (5), (11), (14) and (15) into the static balance Eq. (1) gives:

$$2TA\sin\theta + \gamma_L \pi R_1^2 b \text{ - kt} = \frac{\pi k_1 e^{k_2 t} R_1 \text{b}}{\delta}\frac{dS}{dt} + (G + \delta_v + f - Q_{\max}) \tag{26}$$

To simplify the calculation, let all non-time function parameters be Z, namely:

$$Z = 2TA\sin\theta + \gamma_L \pi R_1^2 b + Q_{\max} - G - \delta_v - f \tag{27}$$

Substituting Eq. (23) into Eq. (22), separating the variables of Eq. (22) and integrating the time parameter (t):

$$\begin{aligned} \int \mathrm{d}S &= \int \frac{\delta e^{-k_2 t}}{\pi R_1 b k_1}(Z - kt)dt \\ S &= \frac{-\delta Z}{\pi R_1 b k_1 k_2} e^{-k_2 t} + \frac{\delta k}{\pi R_1 b k_1}\left(\frac{t}{k_2} e^{-k_2 t} + \frac{1}{k_2^2} e^{-k_2 t} + c_1\right) \end{aligned} \tag{28}$$

Substituting the initial test conditions, when the time t is 0, the segment does not float, and the floating amount S of the segment is 0, then the constant c_1 can be solved:

$$c_1 = \frac{Z}{kk_2} - \frac{1}{k_2^2} \tag{29}$$

Finally, the calculation expression of segment buoyancy is obtained as follows:

$$\begin{aligned} S &= \frac{-\delta Z}{\pi R_1 b k_1 k_2} e^{-k_2 t} \\ &+ \frac{\delta k}{\pi R_1 b k_1}\left[\frac{t}{k_2} e^{-k_2 t} + \frac{1}{k_2^2} e^{-k_2 t} + \frac{Z}{kk_2} - \frac{1}{k_2^2}\right] \end{aligned} \tag{30}$$

4 Application in Practical Project

Taking the shield tunnel between Anfeng station and Dongxin station of Nanchang rail transit line 4 as the project background, the outer diameter of the tunnel is 6 m, the inner diameter is 5.4 m, the segment width is 1.5 m, the shield tail gap is 0.015 m, and the maximum buried depth of the tunnel is 25.8 m. First of all, through the field floating amount monitoring, it is found that during the construction process, the right line 654–683 rings of the section has a relatively serious floating problem. The upward moving of the segment is between 79–96 mm, the upper floating section is composed of silty clay, fine sand, medium sand, coarse sand, gravelly sand, strongly weathered argillaceous siltstone and moderately weathered argillaceous siltstone.

Fig. 7. Field monitoring

Then, the upward moving of segment in this section is calculated theoretically. The reinforced concrete weight value is 25 kN·m^{-3}, the unit weight of grouting slurry is 11.4 kN·m^{-3}, the grouting pressure is 0.28 MPa, the internal friction coefficient is 0.3, the single M24 bolt preload is 3 kN, and there are 17 longitudinal bolts per ring, the radius of bolts is 0.015 m. According to the code for design of steel structures, the allowable shear resistance of M24 bolt is between 367 and 438 kN, the average unit weight of the overlying soil layer is 18.5 kN·m^{-3}, the maximum thrusting force of the jacking cylinder is 39910 kN, the coefficient of lateral earth pressure is 0.8, the internal friction angle is 12°, the cohesive force is 18 kPa, and the stop time of segment floating is about 60 h. In this paper, taking the 654–683 ring as the analysis object, and calculating the force of overlying soil put on segment by the average distribution depth in the segment, the depth is 21 m. Substituting the above parameters into the previous theoretical derivation formula to calculate each parameter and component force, the calculations are as follows: $F_{f1} = 483.25$ kN, $F_{f2} = 2520$ kN, $G = 250.89$ kN, $f = 615.3$ kN, $Q_{max} = 2.40$

kN, $\delta v = 1668.49$ kN, $B_t = 6.29$ m, $S = 87.1$ mm, When considering the degree of soil arching effect, $S = 82.7$ mm, and the calculated average of field monitoring value is 80.9 mm. At last, as shown in Fig. 8. The theoretical calculated value was compared with the measured value of the field monitoring value.

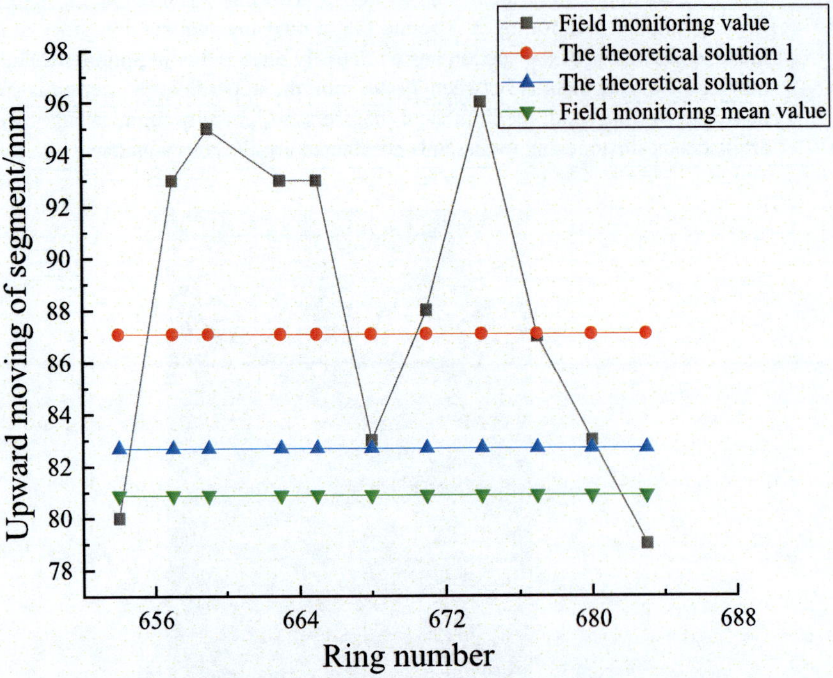

Fig. 8. Comparison between calculated value and measured value of segment floating

5 Effective Anti-floating Measures for Segments on Site

In the previous article, the reasons for the rise of the segment were discussed, and the mathematical calculation model of the floating amount was established based on the stage of the segment floating. In this section, the effective anti-floating measures of the segment were proposed based on the calculation expression. Relying on the tunnel project of the shield tunnel between Anfeng Station and Dongxin Station of Line 4 of Nanchang Rail Transit, after taking measures such as controlling the shield posture, stacking weights in the tunnel, adjusting the value of the tunneling parameters, and adjusting the mixing ratio of the grouting material on the right side of the tunnel, the rise of the segment was controlled to a reasonable extent.

5.1 Controlling the Shield Attitude

It is difficult to avoid the segment floating up in the process of shield tunneling. The excessive serpentine movement of shield machine will cause frequent deviation correction. The most direct impact is that the segment is not evenly stressed. If the stress direction is upward, it will aggravate the segment floating up. Therefore, it is necessary to properly reduce the axis pushed by shield machine and make it do small serpentine movement along the tunnel axis as much as possible [23].

5.2 Stacking Heavy Objects in the Tunnel

A certain weight of heavy block can be placed in the section with serious uplift, which can be seen as the whole segment to provide anti-buoyancy. However, the weight of the heavy block should be within a reasonable range. Too much weight will cause the segment's anti-buoyancy to be greater than the buoyancy it receives, and then the segment will sink. The approximate estimation of the weight of the heavy block can be based on the following formula [21]:

$$M = \mathrm{m}(F_{\mathrm{f}} - G - F_{\mathrm{n}} - f - P, up) \tag{31}$$

Where m is the safety factor and needs to be taken according to the site conditions.

5.3 Adjusting the Value of Tunneling Parameters

In the process of shield tunneling, the rationality of the value of the tunneling parameters will affect the rise of the segment to a certain extent. The main parameters of the tunneling include grouting pressure, shield thrust and tunneling speed, as follows:

1) Grouting pressure is an important factor for the "dynamic buoyancy" of the segment. Increasing the grouting pressure properly is conducive to ensuring the filling degree of slurry in the shield tail gap. However, when excessive grouting pressure is applied, it will cause greater disturbance to the segments and increase the pressure applied on the segments, which will result in the "dynamic buoyancy" of segments [22]. It is found that when the grouting pressure is 0.26–0.29 MPa, it can control the floating of segment effectively;
2) The shield propulsion system adopts 30 jacking cylinders with a cylinder diameter of 220 mm. The jack cylinders are arranged in the form of 8 cylinders at both ends and 7 cylinders at the left and right sides. The thrusting force of the shield is equal to the total thrust provided by the propulsion cylinder. In essence, adjusting the shield thrust is to adjust the jack oil cylinder thrust. Therefore, it is necessary to adjust the jack oil cylinder thrust in each area reasonably to avoid uneven segments caused by excessive local oil pressure, and accelerate segment floating [23];
3) The shield tunneling needs to go through the Ganjiang River, and in most cases, it is conducted in the moderately weathered argillaceous siltstone stratum, which belongs to the hard rock water bearing stratum. When the tunneling speed is too fast, the slurry cannot be condensed in time after grouting behind the wall, and the segment cannot be constrained to float up in a large area effectively. The tunneling speed of the shield should not be greater than 30 mm·min^{-1} and not less than 18 mm·min^{-1}.

5.4 Adjusting the Mixture Ratio of Grouting Slurry

In the whole process of driving, the grouting slurry adopts single liquid slurry. In order to better restrain the segment floating, the grouting slurry should have good slurry performance, that is, the slurry has good filling property, high early strength, appropriate initial setting time and good water resistance dispersion, then slurry and surrounding rock will form an organic whole. The upward moving of the segment can be controlled after the grouting slurry is improved in a certain area, finally the readjusted grouting slurry can be shown in Table 1.

Table 1. Improved ratio of single-liquid slurry

Material	Cement	Fly ash	Calcium-based bentonite	Fine sand	Water	Water glass
Dosage/(kg·m^{-3})	100	350	70	700	480	0.765

After improvement, the initial setting time of grouting slurry is more than 5.5 h, the unit weight is 1140kg·m^{-3}, the viscosity is 17.16 s, and the slurry bleeding rate is 0.05.

6 Conclusions

In this paper, aiming at solving the problem of segment floating of shield tunnel between Anfeng station and Dongxin station of Nanchang metro line 4, the causes of segment floating are analyzed and the calculation model of segment floating which is based on the degree of soil arching effect has been derived by considering grouting construction effect, residual stress of Jack, overburden force, frictional resistance between rings, dead weight of segment and grout and bolt shear resistance. The conclusions are as follows:

1) There is a small deviation between the theoretical calculation result of buoyancy and the measured value of buoyancy. When calculating the theoretical value of segment upward moving by Terzaghi earth pressure theory is 87.1 mm, and the theoretical value of segment upward moving is 82.7 mm which is based on the earth pressure calculation formula proposed by Li Chunlin. It can be found that although the theoretical value calculated based on Terzaghi earth pressure theory is more in the middle of the field monitoring value, it will deviate more from the field monitoring mean value, the theoretical value calculated in this paper is closer to the average value of actual monitoring value.
2) Based on the analysis of the stress components of the segments, a set of effective control measures are put forward. At the construction site, measures such as controlling the posture of the shield, stacking heavy objects in the tunnel, adjusting the tunneling parameters, and improving the performance of the grouting slurry have alleviated the floating problem of the segment to a certain extent. Especially, when the tunnel axis is properly reduced during the segment assembly, the deviation value between the tunnel axis and the tunnel design axis is determined according to the

theoretical value of segment floating, which plays an important role in ensuring the segment health during the assembly, and plays a good role in avoiding the damage of the segment and the segment dislocation.

However, when analyzing the force of grouting on the segment, the slurry is assumed to meet Newton's internal friction law, while the grouting force needs to be corrected for non-Newtonian slurry; The calculation of "dynamic buoyancy" is assumed to be in the form of a fan distribution, when the distribution is triangular, the calculation of "dynamic buoyancy" is worthy of further study. In addition, it is difficult to determine the specific situation of the extent of soil arching effect. In this paper, the situation of soil arching effect when calculating the upward moving of segment is assumed to be a random value, which may be slightly deviated from the actual situation.

References

1. He, C., Feng, K., Fang, Y.: Review and prospects on constructing technologies of metro tunnels using shield tunneling Method. J. Southwest Jiaotong Univ. **50**(01), 97–109 (2015)
2. Ji, C., Zhou, S.-H., Xu, K.: Characteristics of upward moving for lining during shield tunneling construction. Chin. J. Rock Mechan. Eng. **S2**, 3619–3626 (2013)
3. Chen, R.-P., Liu, Y., Liu, S.-X.: Characteristics of upward moving for lining during shield tunneling construction. J. Zhejiang Univ. (Eng. Sci.) **6**, 1068–1074 (2014)
4. Hu, H., Zhang, H., Liu, X.-D.: Analysis on influencing factors of segment floating in construction of shield tunnel in mudstone stratum. Highway. **63**(12), 312–318 (2018)
5. Wang, D.-Y., Yuan, J.-X., Zhu, Z.-G.: Theoretical solution of longitudinal upward movement of underwater shield tunnel and its application. Rock Soil Mech. **35**(11), 3079–3085 (2014)
6. Zhang, J., Zhao, L., Zhou, J.-M.: Research on upward moving mechanism for segment of shield tunnel=. Railway Stand. Design. **60**(10), 88–93 (2016)
7. Li, C.-L.: Method for calculating loosening earth pressure during construction of shield tunnels. Chin. J. Geotechn. Eng.. **9**, 1714–1720 (2014)
8. Lv, G.-Q., Zhou, J.-J., Yang, Z.-X.: Prediction of shield tunnel segment up-floating caused by formation rebound. Tunnel Constr. **37**(A02), 87–93 (2017)
9. Yao, S., Zhou, S.-H., Ji, C.: Analysis of shield tunnel segment uplift data and uplift value forecast during tunnel construction in variable composite formation. Chin. J. Rock Mech. Eng. **36**(S1), 3464–3474 (2017)
10. Ye, F., Zhu, H.-H., Ding, W.-Q.: Analysis on anti-buoyancy calculation in excavation of big cross-section shield tunnel. Chin. J. Underground Space Eng. **3**(5), 849–853 (2007)
11. Yuan, W., Liu, S.-N., Wang, L.: Study on the segment upward moving of large diameter slurry shield tunnel after the initial setting of grouting. Concrete **1**, 131–135 (2020)
12. Zhang, J., Zhao, L., Zhou, J.-M.: Research on upward moving mechanism for segment of shield tunnel. Railway Stand. Design. **60**(10), 88–93 (2016)
13. Ye, F.: Analysis and control of floating mechanism during construction of soft soil shield tunnel. Tongji University (2007)
14. Ye, F., Zhu, H.-H., He, C.: The grouting diffusion mode behind the shield tunnel wall and the pressure analysis of the segment. Rock Soil Mech. **5**, 1307–1312 (2009)
15. Song, K.-Z., Yuan, D.-J., Wang, M.-S.: Back- filled grouts diffusion model and its pressure to segments of shield tunnel. Rock Soil Mech. **29**(3), 619–623 (2008)
16. Li, Y.-L.: Research on the mechanical characteristics of segment structure during shield tunnel construction. Beijing Jiaotong University (2008)

17. Terzaghi, K.: Theoretical Soil Mechanics, vol. 15, pp. 37–42. John Wiley and Sons, Inc., New York (1943)
18. Lou, P.-J., Xu, Y.: Discussion on "method for calculating loosening earth pressure during construction of shield tunnels." Chin. J. Geotech. Eng. **37**(7), 1353–1354 (2015)
19. Kezdi, A.: Lateral earth pressure. In: Winterkorn, H.F., Fang, H.Y. (eds.) Foundation Engineering Handbook, pp. 147–220. Van Nostrand Reinhold Co., New York (1986)
20. Xu, C.-J., Liang, L.-J., Chen, Q.-Z., et al.: Research on loosening earth pressure considering the patterns of stress distribution in loosening zone. Rock Soil Mech. **39**(06), 1927–1934 (2018)
21. Dai, Z.-R.: The Mechanism and control principle of upward movements of segments at the rear of shield tail. China Railway Science. **34**(01), 59–66 (2013)
22. Liu, H.-Q.: Control measures and application research of floating segments of water-rich hard rock section in a shield tunnel. Huazhong University of Science and Technology (2019)
23. Dong, S.-S., Yang, P., Jiang, C.-Y.: Analysis of mechanism and controls of segment floating of shield tunnels. Chin. J. Underground Space Eng. **12**(01), 49–54 (2016)
24. Xu, W.-Z., Liu, S.-J., Liao, S.-M.: Analysis on the influence of shield depth on soft soil arching effect. Chin. J. Underground Space Eng. **13**(S1), 65–69 (2017)

Invert Heave Disease and Treatment Measures of Operating Railway Tunnel in Horizontal Layered Rock Mass

Linyi Li[1](✉), Junsheng Yang[1], Jian Wu[2], Shuying Wang[1], Xinghua Fang[1], and Maolong Xiang[1]

[1] School of Civil Engineering, Central South University, Changsha, Hunan, China
tunnel_lly@csu.edu.cn

[2] China Railway Southwest Research Institute Co., Ltd., Chengdu, China

Abstract. During operation of the Yunding Tunnel, the uplift displacement of the invert increases continuously, seriously impacting the safety of the tunnel structure and the normal operation of the railway line. To study the occurrence mechanism and evolution law of invert heave disease of railway tunnel in horizontal layered rock mass, both field investigation and numerical simulations were carried out, and the influence of layer thickness of layered rock mass and horizontal in-situ stresses were discussed. The results show that the deformation characteristic of layered rock mass is the main factor leading to the disease in operating tunnels. Especially in thin-thick to medium-thick layered rock mass and the horizontal in-situ stress of K_0 less than 1.0 or more than 2.0, it is difficult to ensure the stability and safety of bottom structure. Moreover, two proposed treatment schemes were analyzed to evaluate the regulation effect, and the latter scheme was recommended to be used in the project. This study can provide a reference for the treatment of operating tunnel in horizontal layered rock masses.

1 Introduction

Layered rock masses, whose key characteristic is the existence of layered structural planes, are widely distributed all over the world. Unlike intact rock, the behavior of layered rock masses depends not only on the properties of the intact rock, but also on the structural planes between layered rock mass (Adhikary and Dyskin 1997). Due to these structural planes, such rock masses have heterogeneity and anisotropy (e.g., deformation and strength). Excavation in layered rock masses is a process in which the stress field of surrounding rock changes constantly (Fortsakis et al. 2012). Such changes during unloading can cause the layered rock masses to slide, separate, and flex along structural plane (Zhou et al. 2019). This will further lead to the change of rock mass structure and a deterioration of its mechanical properties. Especially if there is high in-situ stress, various forms of problem may occur, including spalling and rheological behavior of surrounding rock, and cracking and large deformation of support (Tang et al. 2016). However, many

J. J. Yang et al. (Eds.): GeoChina 2021, SUCI, pp. 37–56, 2021.
https://doi.org/10.1007/978-3-030-79672-3_3

mountain tunnels must be or have been built in the layered rock masses as the wide distribution of these formation. Thus, many researchers studied the effect of layered rock masses on the stability of tunnel engineering during excavation, including analysis, numerical simulations, field tests and physical test (Jia and Tang 2008; Małkowski 2015; Yang et al. 2018; Do et al. 2019; Moussaei et al. 2019). Based on those researches, practical studies on how to ensure the stability and safety of underground engineering during excavation were carried out, and some useful strategies have been proposed, such as deformation prediction (Li et al. 2020) and support design optimization (Ding et al. 2019).

Railway tunnels, as the important part of mountain tunnels, carry freight transport or passenger transport railway lines. Due to the strict standard of operating tunnel on track deformation, railway tunnel is often concerned with its structural stability of tunnel bottom during operation, and even a small displacement of the railway bed can cause terrible disaster (Li et al. 2018; Qian et al. 2019). However, because of the complexity of rock mechanical properties and deformation characteristics, the stability of tunnel bottom structure is difficult to be guaranteed in layered rock masses, especially for horizontal layered rock (Jia and Tang 2008; Wang et al. 2014). Even after the completion of tunnel for many years, the unstable displacement of tunnel bottom structure still exists. For instance, the Yunding tunnel, constructed in 2009, carries a railway line. Its design speed of tunnel operation is 200 km/h. During the tunnel construction and initial operation, there was no obvious stability problem. However, since May 2012, an uplift displacement appeared and developed continuously in the tunnel invert. The unstable displacement not only lead to the cracking of bottom structure, but also caused the train speed must to be limited to less than 80 km/h up to now. Meanwhile, some similar diseases are also found in other railway tunnels as shown in Table 1, which seriously disturbed the normal operation of railway line. Unfortunately, few studies focus on the effect of horizontal layered rock mass on the stability of operating tunnel, and the evolution law of invert heave disease is still unclear.

In this paper, the performance of an operating railway tunnel in horizontal layered rock masses was evaluated by field investigation and numerical analysis. Geology factors, such as rock mass structure, in-situ stress condition were surveyed by field investigation, whereas a numerical simulation considering the mechanical properties of layered rock mass was carried out to discuss how rock mass structure and in-situ stress condition affect the deformation law of tunnel structure and to evaluate the safety of tunnel support. Moreover, the proposed two treatment schemes were analyzed to evaluate the suitability for tunnel renovation. The results of numerical analysis are in good agreement with the field data, which was expected to provide a reference for the disease treatment and design of railway tunnel in horizontal layered rock masses.

Table 1. Typical cases of structural instability of operating railway tunnels in layered rock mass.

No	Tunnel	Design speed	Geological condition of disease segment	Disease characteristics	Adverse effects
1	Guangzhao (Xu 2019)	350 km/h	With a buried depth of about 300 m, the lithology of dolomite and mudstone, a dip angle of the bedding plane of 5° to 15° and medium-thick layered rock mass structure	Invert and track slab heave; Cracking in track slab and invert filling	Train speed limited to 60 km/h for three years
2	Gangwu (Li 2017)		With a buried depth of about 450 m, the lithology of limestone and mudstone, a dip angle of the bedding plane of 10° to 20° and medium-thick layered rock mass structure	Invert and track slab heave; Cracking in invert filling	Train speed limited to 80 km/h for two years
3	Zhaojiayan (Jia 2018)	250 km/h	With a buried depth of about 500 m, the lithology of limestone and mudstone, a dip angle of the bedding plane of 15° and medium-thick layered rock mass structure	Invert and track slab heave; Cracking in track slab	Train speed limited to 80 km/h
4	Gaopo (Zhang and Ding 2020)		With a buried depth of about 400 m, the lithology of sandstone and mudstone, a dip angle of the bedding plane of less than 5° and thin-layered rock mass structure	Invert and track slab heave; Cracking in track slab and invert filling	High cost for tunnel renovation
5	Yunding	200 km/h	With a buried depth of about 180 m, the lithology of sandstone and mudstone, a dip angle of the bedding plane of less than 5° and medium-thick layered rock mass structure	Invert and track slab heave; Cracking in track slab and invert filling	Train speed limited to less than 80 km/h
6	Maotianshan (Su 2010)		With a buried depth of about 150 m, the lithology of sandstone and shale, a dip angle of the bedding plane of 5° to 10° and thin-thick layered rock mass structure	Invert and track slab heave; Cracking in invert filling; separation of track slab and invert filling	Train speed limited to 60 km/h

Note: According to the Chinese Code for investigation of geotechnical engineering (GB50021-2018), rock mass structure can be determined by layer thickness (L). thin-layer represents "$L <$ 0.1 m"; medium-thick layer represents "0.1 m $< L <$ 0.5 m"; thick layer represents "0.5 m $< L$ $<$ 1.0 m"; extremely-thick layer represents "$L >$ 1.0 m"

2 Project Overview

A series of field investigations were carried out to research the operating Yunding tunnel with invert heave disease which passes through a formation with layered rock masses. The Yunding tunnel of Dazhou-Chengdu railway is located in Sichuan Province, China.

The nearly 7.86 km long tunnel, constructed in June 2012, carries a double-track mixed passenger and freight railway line. Since May 2012, the phenomenon of structural instability at tunnel bottom was observed during tunnel operation. The detection results of track deformation shown that there was an uplift displacement located in the section of K108+600 ~ +700. Then, a regular monitoring of structural displacement was carried out every half a month. The monitoring results of K108+620 was shown in Fig. 1, and it was obvious that the uplift displacement of tunnel bottom was increasing. By January 2014, the displacement of track slab was up to 22.7 mm. Moreover, some cracks were also found on the surface of invert filling and track slab, as shown in Fig. 2.

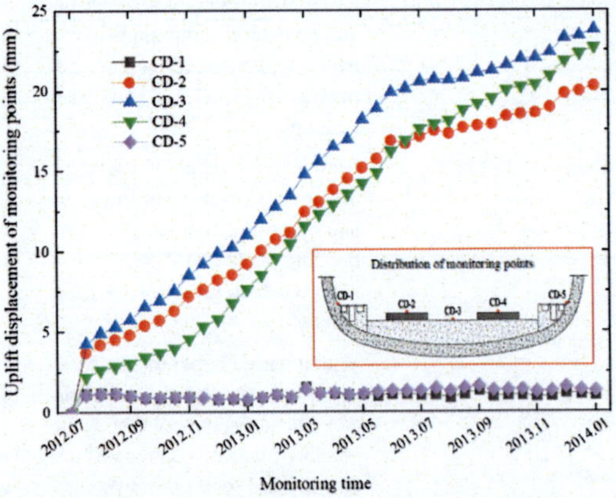

Fig. 1. Displacement development curve of measuring points at the section of K108+600.

Fig. 2. Cracking of invert filling and track slab.

The cover depth of the studied sections (K108+600 ~ +700) was approximately 180 m. The surrounding rocks in the studied sections was specified as Grade III according to Chinese Code for Design of Railway Tunnel (TB10003-2016), which means that the rock had a complete rock mass. It is a NATM tunnel. As shown in Fig. 3, the tunnel was supported by composite lining, including primary support, secondary lining and invert filling. The width and height of tunnel were 13.30 m and 11.77 m respectively, whose span ratio (height:width) was approximately 1.13. During tunnel construction, there was no structural stability problem in the studied sections and not much groundwater was encountered.

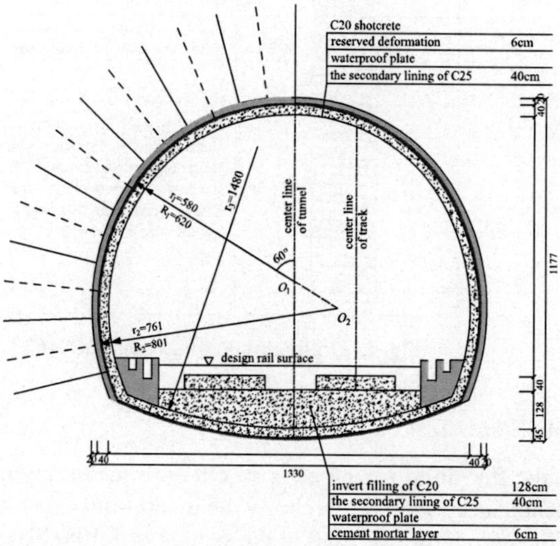

Fig. 3. Cross section of Yunding tunnel (Grade III, unit: cm).

3 Field Investigation of Geological Conditions and Tunnel Support

3.1 Survey of Lithology and Rock Mass Structure

After invert heave disease occurred in the tunnel, to further analysis the causes of uplift displacement, a core drilling survey of rock mass at the bottom of tunnel was carried out. From October to November 2012, a total of 16 boreholes were drilled in the study section, with a maximum drilling depth of 8.5 m. Most of boreholes had similar results, and the typical drilling result of section K108+620 was shown in Fig. 4. It can be seen from Fig. 4 that: for the depth 0–1.6 m, it was concrete with a good compactness; for the depth 1.6–1.8 m, it was strongly weathered mudstone, with a low strength; for the depth 1.8–8.5 m, it was medium-thick layer mudstone interlayered with medium-thick layer sandstone and sandy mudstone, with a dip angle of less than 5°. It is obvious that the surrounding rock of tunnel bottom had the characteristics of layered rock mass. The existence of broken rock layer of depth 1.6–1.8 m might be related to the dynamic load of train during tunnel operating. Moreover, seven core samples from drilling were test for its expansibility. The test results shown that the free swelling rate of all rock samples were less than 22%. According to Chinese Code for Geology Investigation of Railway Engineering (TB10012-2019), the swelling of surrounding rock has little effect on engineering structure, and the structural instability can be excluded as a result of swelling of rock mass.

Depth	Material type	Lithology and rock mass structure
0~1.6 m	Contrete	Contrete with a good compactness, its aggregate with a good uniformity
1.6~1.8 m	Rock	Rock was broken and massive mudstone with a low strength
1.8~8.5 m	Rock	Medium-thick layer mudstone interlayered with medium-thick sandstone and sandy mudstone. Rock was moderately weathered. The layered plane was near horizontal with a dip angle of less than 5°

Fig. 4. Drilling result of section K108+620.

3.2 In-Situ Stress Test

The hydraulic fracturing technique is an effective measuring method in the geostress measurement in deep stratum. To clarify the in-situ stress conditions in studied section, hydraulic fracturing test was used in the section of K108+620, and the test result was shown in Table 2. As we can see from Table 2, the maximum horizontal stress is greater than the vertical stress, and the lateral pressure coefficient (σ_H:σ_v) at different depths ranges from 1.28 to 1.81. It is obvious that the in-situ stress in the studied section is dominated by tectonic stress. Moreover, considering the effect of unloading, the stress measured at the depth of 5.9 m was obviously less than its original in-situ stress, which is also the reason why the lateral pressure coefficient of 5.9m depth is different from other depths.

3.3 Inspection of Tunnel Support

For inspecting construction quality of tunnel support, according to Chinese Technical Specification for Testing Concrete Strength with Drilled Core Method (JGJ/T 384-2016), seven drilling holes at bottom structure were set in the studied section, and the compressive strength of all core samples was test. The inspection results showed that the thickness of concrete structure (including invert and invert filling) meets the design requirement and the concrete strength also meet the standard requirement (TB10003-2016). Thus, it is concluded that the structural instability of Yunding Tunnel should not be related to the construction quality.

4 Three-Dimensional Numerical Analysis

4.1 Numerical Modeling with Layered Rock Mass

Numerical Analysis was carried out to study the effect of layered rock mass on the deformation and structural safety of the existing tunnel. Five kinds of three-dimensional numerical model were established using FLAC3D software, and for Case 2–5, a further discussion on the stability of bottom structure under the influence of rock mass structure and in-situ stress condition was carried out, as shown in Table 3. The varied situations of those cases were expected to explain how the layered rock mass affect the deformation and stress state of the concrete of bottom structure.

Table 2. Test result of in-situ stress in the section of K108+620.

Depth/m	σ_H/MPa	σ_h/MPa	σ_v/MPa	Direction angle of σ_H
5.9	6.87	3.73	5.35	N5°W
7.8	9.67	5.20	5.41	N6°W
9.9	9.63	5.04	5.46	N8°W
11.9	9.98	5.49	5.51	N8°W

Note: σ_H is the horizontal maximum principal stress; σ_h is the minimum horizontal principal stress; σ_v is the vertical stress

Table 3. Numerical model and the studied cases

Case	Situation	Layer thickness of rock
1	Homogeneous rock	–
2	Thin-thick layered rock	0.2 m
3	Medium-thick layered rock	0.5 m
4	Thick layered rock	1.0 m
5	Extremely-thick layered rock	1.5 m

To minimize the boundary effect, the distance between center of the tunnel and the boundaries of Z and X directions is more than 5 times the tunnel span in this analysis (Möller 2006). The cover depth of the studied section is approximately 180 m, and the geostatic stress of overlying rock mass was applied on the top of the model. In order to simplify the calculation of Case 2–5, the area of layered rock mass was only set at the scope of 2 times the tunnel span around the tunnel. The homogeneous rock and layered rock were modeled with the Mohr-Coulomb elastoplastic model. As shown in Fig. 6, interface elements have the properties of friction, cohesion, normal and shear stiffness, and tensile and shear bond strength and are characterized by Coulomb sliding or tensile and shear bonding. With the interface, the slide or separation of discontinuities can be realized in FLAC3D simulation (ICGI 2012). Thus, interface elements were set between the adjacent layered rock masses to simulation the interlayer discontinuity (see Fig. 5). According to the support scheme of tunnel, the structure mesh inside the tunnel was established finely, considering primary support, secondary lining, invert filling and track slab. All support structures were simulated as linear isotropic elastic materials. Figure 5 shows the typical numerical model of Case 3 and its boundary condition. Properties of rock mass, support structure and the discontinuities from the results of geology investigation were described in Tables 4 and 5, respectively.

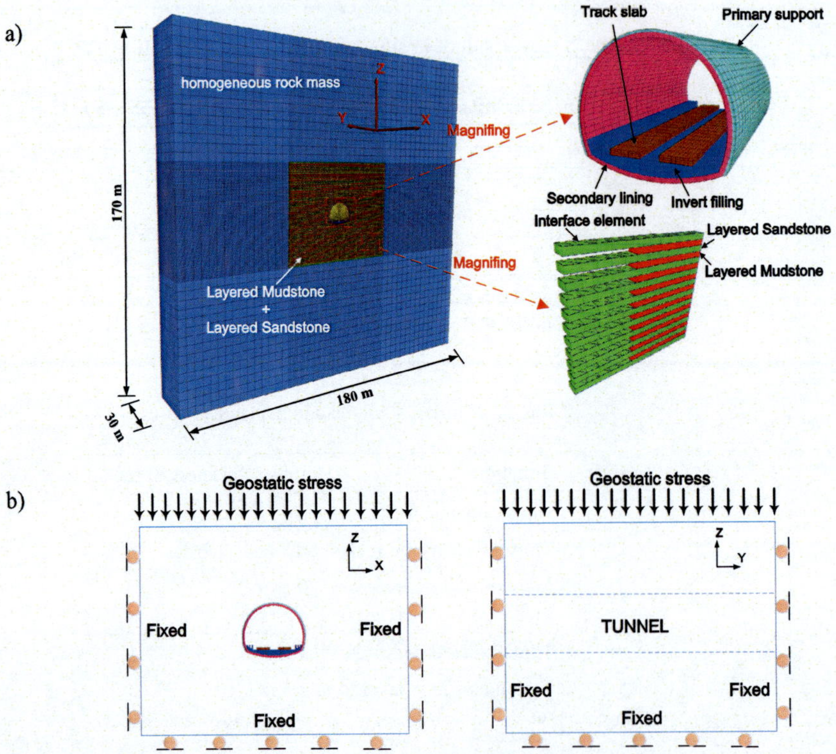

Fig. 5. Numerical model for railway tunnel with layered rock mass. (a) Meshing and size of model; (b) Boundary condition.

4.2 Comparison for Cases with and Without Layered Rock Mass

The behavior of the tunnel structure resulted in different characteristics when the surrounding rock with or without layered rock mass. According to the investigation results of in-situ stress (Table 2), lateral pressure coefficient in numerical simulation was set as 1.75. Six model were calculated, including homogeneous rock of mudstone, homogeneous rock of sandstone, and four thickness cases of layered rock mass for Case 2–5. The vertical displacement distributions of bottom structure are presented in Fig. 7, and the location of monitoring points were shown in Fig. 1. It is obvious that different degree of uplift displacement has occurred in tunnel bottom structure. As shown in Fig. 7(a), the distribution of uplift displacement is characterized by "large in the middle and small

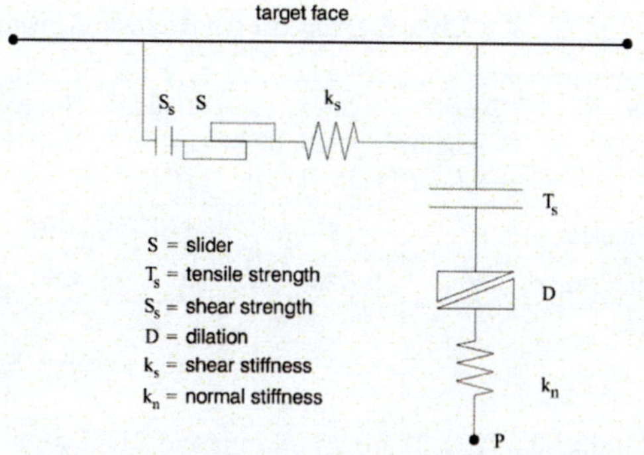

Fig. 6. Components of the interface element.

Table 4. Properties of rock mass and support structure for the studied section

Material	Density [γ(kN·m^{-3})]	Elastic modulus [E(GPa)]	Poisson's ratio	Cohesive strength [c(kPa)]	Friction angle [φ(°)]	Compressive strength [σ(MPa)]
Homogeneous rock	23.1	5.80	0.28	2.20	42	11.88
Mudstone	22.2	3.20	0.32	1.42	39	5.31
Sandstone	24.0	8.40	0.26	2.98	45	18.45
Mudstone with grouting	23.2	5.40	0.28	2.00	42	8.00
Sandstone with grouting	24.5	9.60	0.24	3.50	48	22.50
Primary support	23.0	25.00	0.20	–	–	–
Secondary lining	24.0	30.00	0.20	–	–	–
Invert filling	23.4	28.00	0.20	–	–	–
Track slab	25.0	33.50	0.20	–	–	–

on both side". Meanwhile, the rock structure has influence on the uplift displacement. It is interesting that the uplift displacement of bottom structure in layered rock mass is mostly larger than that of homogeneous rock mass. The results show that layered rock mass easily causes the instability of tunnel structure, and in layered rock mass, the tunnel invert is prone to large uplift deformation. For layered rock mass, layer thickness also affects the stability of tunnel bottom. With the increase of layer thickness, the uplift displacement of tunnel invert decreases obviously, which means that a small layer thickness is unfavorable to the stability of tunnel invert. As shown in Fig. 7(b), when railway tunnel is located in medium-thick layered rock mass, the uplift displacements of invert filling

Table 5. Properties of interface element.

Discontinuity	Tensile strength [T_s(kPa)]	Shear strength [S_s(kPa)]	Friction angle [φ(°)]	Normal stiffness [k_n(GPa·m^{-1})]	Shear stiffness [k_s(GPa·m^{-1})]
Sandstone-mudstone	80	80	25	3.50	1.25
Sandstone-mudstone with grouting	420	420	32	4.50	2.00

(CD-3) and track slab (CD-2) are 25.88 mm and 21.55 mm, respectively. As expected, the results of numerical analysis were in good agreement with the field monitoring data, which clearly demonstrated that layered rock mass structure is the main reason for the invert heave disease of Yunding tunnel.

The numerically computed stresses of tunnel structure at typical locations are presented in Table 6 to illustrate the influences of layered rock mass on the stress state of tunnel support. As shown in Table 6, the stresses of lining structure were always in compressive state, while at some typical locations the stresses of invert filling and track slab were in tensile state. The maximum compressive strength and tensile stress for C20, C25 and C40 concrete are 1.7 and −15.5 MPa, 2.0 and −19.0 MPa, and 2.7 and −29.5 MPa, respectively (TB 10003-2016). It is obvious that those bold values in Table 6 mean that the stress values exceeded the limit, and the concrete at these locations was cracked. It is concluded that the uplift of bottom structure in layered rock mass causes the increase of the structural tensile stress, and with the decrease of layer thickness, the tensile stress of bottom structure increases significantly. Figure 8 shows the distribution of the maximum principal stress (σ_1) of medium-thick layered rock (Case 3). The structural tensile area was concentrated in invert filling and track slab, and based on numerically computed values, concrete should be cracked on the top surface of track slab and invert filling. Such results were consistent with the crack distribution observed in the field (Fig. 2). Therefore, the layered rock mass can cause the instability of tunnel bottom structure, including structure uplift and the increase of tensile stress. Especially in thin-thick to medium-thick layered rock mass, such instability of tunnel bottom structure is the most serious and should attract the greatest attention.

4.3 Influence of Horizontal In-Situ Stress

An important role of tunnel support is to restrain the deformation of surrounding rock and maintain the stability of tunnel. The behavior of tunnel support results in different characteristics when the tunnel is under different lateral pressure conditions. A model of tunnel with different lateral pressure coefficient (K_0) was analyzed to discuss the effect of horizontal in-situ stress on the behavior of tunnel structure. The values of K_0 range from 0.5 to 4.0, including 0.5, 0.75, 1.0, 1.25, 1.50, 1.75, 2.0, 2.5, 3.0 and 4.0. Figure 9 shows the variation of vertical displacement at measuring point of CD-2 (track slab) with the increase of K_0. It is interesting that with the increase of K_0, the uplift displacement of tunnel in layered rock mass decreases first and then increases, while

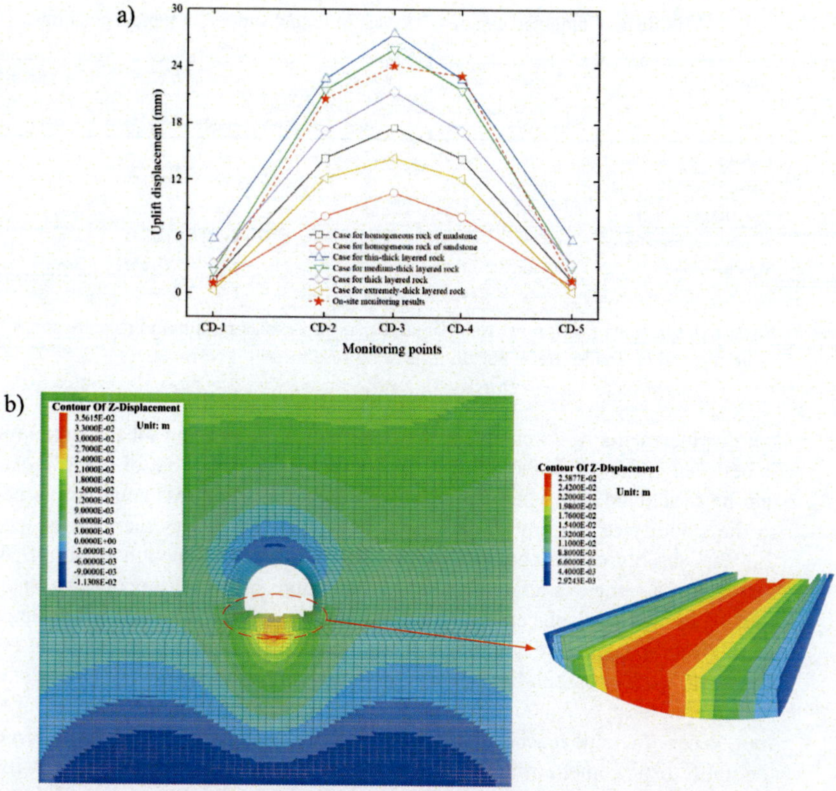

Fig. 7. Distribution of vertical displacement. (a) Curve of uplift displacement of measuring points at tunnel bottom. (b) Vertical displacement of typical case for medium-thick layered rock.

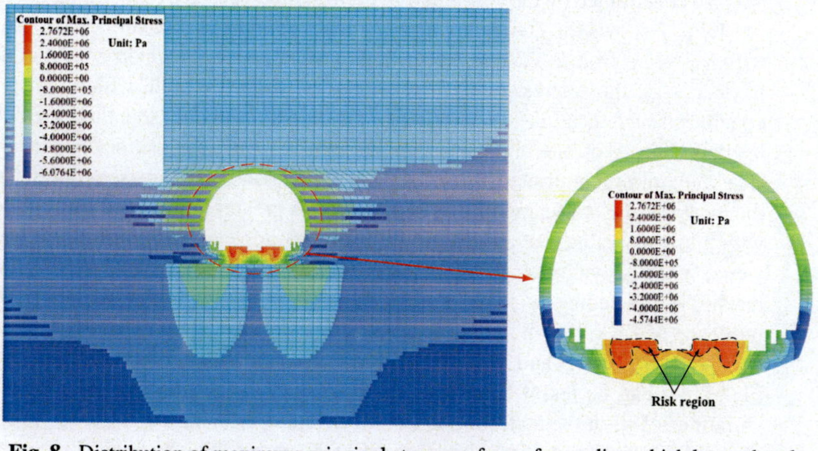

Fig. 8. Distribution of maximum principal stresses of case for medium-thick layered rock.

Table 6. Computed concrete stresses of tunnel support at typical positions.

Cases	Lining of vault (C25)		Lining of side wall (C25)		Invert filling (C20)		Track slab (C40)	
	σ_1 (MPa)	σ_2 (MPa)	σ_1 (MPa)	σ_2 (MPa)	σ_1 (MPa)	σ_2 (MPa)	σ_1 (MPa)	σ_2 (MPa)
Homogeneous rock of mudstone	−1.02	−13.21	−3.97	−13.42	1.49	−5.76	1.83	−4.07
Homogeneous rock of sandstone	−0.66	−8.03	−3.54	−8.92	0.88	−4.21	0.92	−2.27
Thin-thick layered rock	−1.85	−11.48	−4.87	−14.86	**2.08**	−6.67	**2.84**	−5.73
Medium-thick layered rock	−1.65	−16.33	−4.57	−15.53	**1.92**	−6.32	**2.76**	−5.62
Thick layered rock	−1.36	−11.02	−4.36	−14.12	1.65	−6.02	2.06	−3.68
Extremely-thick layered rock	−0.88	−9.82	−2.87	−10.33	1.01	−4.53	1.36	−3.14

Note: σ_1 represents maximum principal stress; σ_2 represents minimum principal stress; positive value represents tensile stress; negative value represents compressive stress

the displacements in homogeneous rock is almost decreasing all the time. When K_0 is between 1.0 and 1.5, the uplift displacement is less than that of other K_0 values. It can be concluded that when K_0 is 1.0 to 1.5, the in-situ stress condition is beneficial to the stability of the bottom structure, and when K_0 is less than 1.0 or more than 2.0, the stability of the bottom structure should attract an enough attention. With the increase of K_0, the overall deformation distributions of tunnel support also change, and Fig. 10 shows the displacement distribution of typical K_0 values (0.5, 1.75 and 3.0) in medium-thick layered rock masses. It can be found that tunnel deformation is mainly characterized by vertical displacement of invert and vault when K_0 value is small, and with the increase of K_0, an obvious convergence deformation also appears at the arch foot. When K_0 value reaches 3.0, the horizontal convergence of arch foot even exceeds the uplift displacement of bottom structure. This means that for the tunnel with a high K_0, the structural deformation of the arch foot should also be concerned. In addition, such results also show that with the increase of K_0, the behavior of bottom structure is not only affected by the warping deformation of layered rock masses below the invert, but also constrained by the compression deformation caused by horizontal in-situ stress.

Table 7 shows the concrete stress of tunnel support at typical locations for Case 3 with K_0 values from 0.5 to 4.0. According to the strength of support concrete (described in Sect. 4.2), those stress values exceeding the limit are bold, which means that the concrete at such locations was cracked or crushed. It is obvious that there are differences in the influence of lateral pressure coefficient on different position. For the vault and side wall, minimum principal stress (σ_2) of lining concrete increases continuously with the increase of K_0. Meanwhile, σ_2 of side wall is greater than that of vault under tunnel with a low K_0 value, but an opposite result of σ_2 appears under tunnel with a high K_0 value. When K_0 value exceeds 3.0, the concrete of vault is crushed, and when K_0 value reaches 4.0 the concrete of vault and side wall are both crushed. To clearly reveal the influence of K_0 value on the cracking of tunnel support, the σ_1 distributions of typical K_0 values (0.5, 1.75 and 3.0) in medium-thick layered rock masses are presented in Fig. 11. It can be learnt from Fig. 11 and Table 7 that with the increase of K_0, the variations of σ_1 at the arch foot, invert filling and track slab are different. For lining of arch foot, σ_1 increases with the increase of K_0 value, which should be caused by the growth of horizontal convergence. For tunnel bottom structure, when K_0 value is low, its behavior is dominated by the warping deformation of layered rock masses below the invert (Fig. 12(a)), which results in a high tensile stress and cracking risk on the surface of invert filling and track slab. With the increase of K_0, the horizontal compression deformation caused by horizontal in-situ stress restrains the uplift of bottom structure (Fig. 12(b)), which results in the redistribution of the tensile stress in bottom structure. Finally, as K_0 value increases, the σ_1 of track slab decreases and the cracking risk region of invert filling shifts from the surface to a deeper position, which might result in the separation of track slab from invert filling.

Generally, for tunnel with a low K_0 value, more attention should be paid to the cracking risk and uplift displacement of bottom structure, and other risks need to draw the attention of engineers with the increase of K_0, including the cracking of arch foot, the crushing of vault and side wall, and the separation of track plate and invert filling.

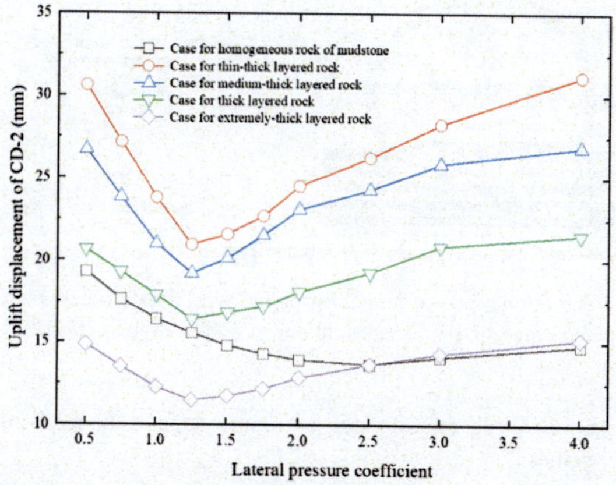

Fig. 9. Uplift displacement of measuring point CD-2 with different K_0 values.

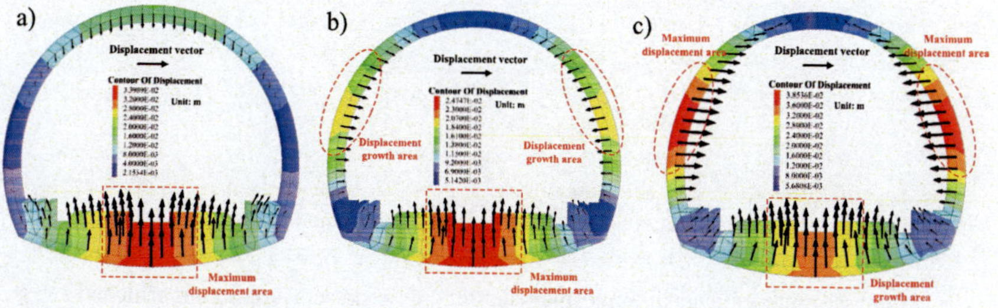

Fig. 10. Distribution of structure displacement with different K_0 values. (a) $K_0 = 0.5$. (b) $K_0 = 1.75$. (c) $K_0 = 3.0$.

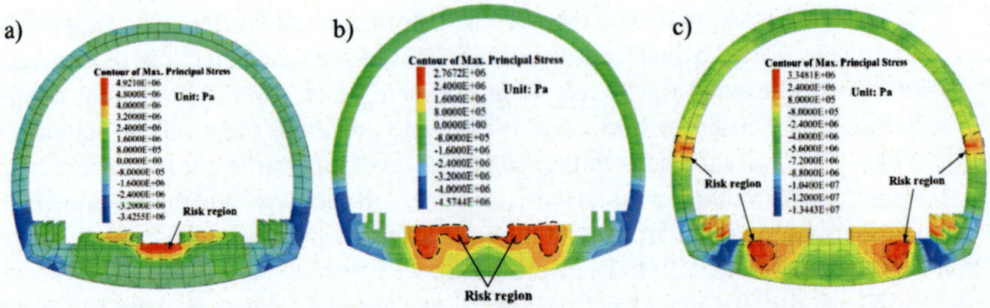

Fig. 11. Distribution of structural maximum principal stresses with different K_0 values. (a) $K_0 = 0.5$. (b) $K_0 = 1.75$. (c) $K_0 = 3.0$.

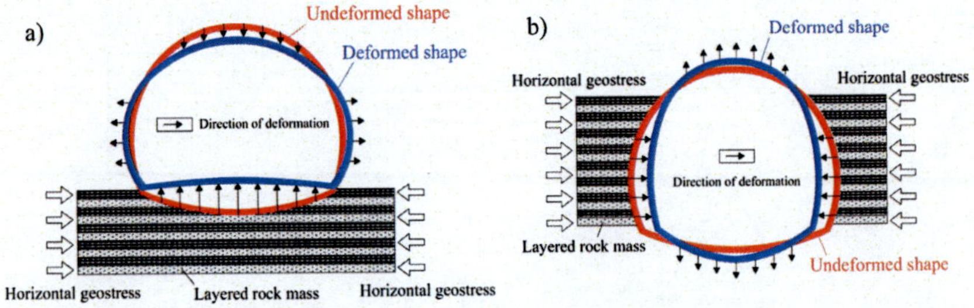

Fig. 12. Deformation mode of tunnel lining in layered rock masses with horizontal in-situ stress. (a) Warping deformation. (b) Horizontal compression deformation.

Table 7. Computed concrete stresses of tunnel support in medium-thick layered rock with different K_0 values.

Lateral pressure coefficient	Lining of vault (C25)		Lining of arch foot (C25)		Lining of side wall (C25)		Invert filling (C20)		Track slab (C40)	
	σ_1 (MPa)	σ_2 (MPa)	σ_1 (MPa)	σ_2 (MPa)	σ_1 (MPa)	σ_2 (MPa)	σ_1 (MPa)	σ_2 (MPa)	σ_1 (MPa)	σ_2 (MPa)
$K_0 = 0.5$	−0.42	−5.12	−1.22	−9.42	−3.42	−13.23	**4.92**	−2.63	**4.01**	−2.19
$K_0 = 1.0$	−0.73	−9.22	−0.65	−9.31	−3.59	−13.89	**3.34**	−3.42	**3.61**	−2.88
$K_0 = 1.25$	−0.89	−11.12	−0.21	−9.22	−3.94	−14.13	**2.63**	−4.05	**3.36**	−3.32
$K_0 = 1.5$	−1.31	−13.43	0.11	−9.03	−4.12	−14.89	**2.16**	−5.34	**3.02**	−4.01
$K_0 = 1.75$	−1.65	−16.33	0.47	−8.95	−4.57	−15.53	**1.92**	−6.32	**2.76**	−5.62
$K_0 = 2.0$	−1.81	−17.82	0.89	−8.90	−4.76	−16.83	**2.28**	−7.05	2.57	−6.23
$K_0 = 3.0$	−2.52	**−21.55**	1.52	−8.83	−5.78	−18.41	**2.88**	−9.42	2.22	−7.92
$K_0 = 4.0$	−3.49	**−25.42**	**2.21**	−8.78	−7.89	**−21.33**	**3.34**	−13.44	1.53	−9.34

Note: σ_1 represents maximum principal stress; σ_2 represents minimum principal stress; positive value represents tensile stress; negative value represents compressive stress.

5 Numerical Analysis of Two Treatment Schemes

5.1 Options for Treatment Scheme

In order to ensure the stability of the bottom structure in the follow-up operation of Yunding Tunnel, two different treatment schemes, namely, structure anchoring scheme (Scheme 1) and reinforcement scheme for surrounding rock at bottom (Scheme 2), are considered as shown in Fig. 13. For the Scheme 1, its principle is to anchor the bottom structure with the deep surrounding rock, so as to limit the uplift displacement. Specifically, the prestressed cables with a length of 20 m and a pretension of 450 kN were adopted and set at the invert filling (Fig. 13(a)). The diameter of prestressed cables is 130 mm and its spacing is 450 m. Moreover, long bolts were also adopted and set at track slab, with a length of 10 m, a diameter of 110 mm, a longitudinal spacing of 1.5 m and a pretension of 80 kN. For the Scheme 2, some grouting measures were adopted to strengthen the mechanical properties of surrounding rock below the invert and weaken the layered characteristics of rock masses. Specifically, as shown in Fig. 13(b), seven grouting bolts and two steel pipe piles were installed at the bottom structure, with a length of 7 m and a longitudinal spacing of 1.5 m. The diameter of bolt and steel pipe pile is 110 mm and 127 mm, respectively. The grouting material was selected as cement slurry with the water-cement ratio of 0.38 to 0.45, content of fly ash and early strength agent of 25% and 3%. During grouting, the grouting pressure should be limited to 1.0 MPa to prevent the uplift of bottom structure.

A numerical model of tunnel with treatment measures in medium-thick layered rock (Case 3) and in-situ stress condition of $K_0 = 1.75$ was built to evaluate the effect of the above two treatment schemes. For the Scheme 1, the prestressed cables and bolt were simulated by cable element in FLAC3D, and for the Scheme 2, the grouting bolt and steel pipe pile were simulated by cable element and pile element, respectively. The physical and mechanical parameters of all structure elements are listed in Table 8. Moreover, considering the reinforcement of layered rock by grouting, the mechanical parameters of mudstone, sandstone and their structural plane were strengthened, and those parameters is listed in Table 4 and 5.

5.2 Displacement of Tunnel Structure

The displacement distribution of tunnel support with or without treatment measures was recorded, and the analysis results are presented in Fig. 14 and 15. It is obvious that after tunnel renovation, the uplift displacement of bottom structure is smaller. Such results show that both the two treatment schemes have some beneficial effects on control the stability of bottom structure. However, some differences of effect on restricting structural displacement still exist between the two treatment schemes. With Scheme 1, the uplift displacement of invert filling (CD-3) and track slab (CD-2) is decreased from 25.88 and 21.55 mm to 15.21 and 12.56 mm, while that of Scheme 2 is reduced to 6.32 and 5.98 mm. It can be concluded that the effect of Scheme 2 is better than that of Scheme 1. Moreover, the results also show that the two schemes can restrain the warping deformation of layered rock masses at the bottom, but have little effect on horizontal compression deformation.

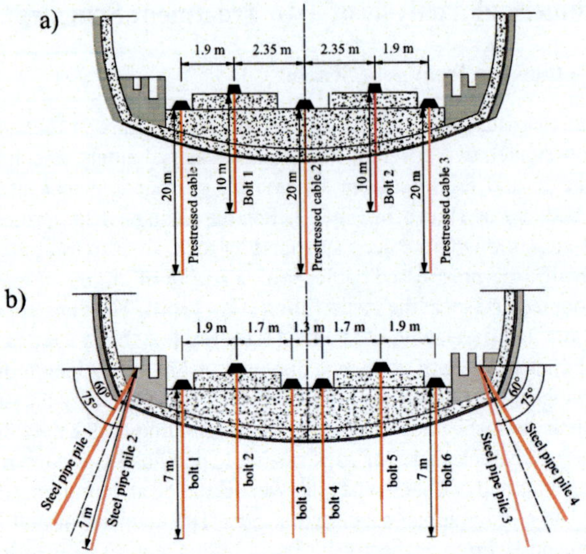

Fig. 13. Cross section of options for the treatment scheme (a) Scheme 1. (b) Scheme 2.

Table 8. Properties of structure elements for tunnel renovation

Type	Diameter [D(mm)]	Elastic modulus [E(GPa)]	Poisson's ratio	Pretension [F(kN)]	Normal stiffness [k_n(GPa·m^{-1})]	Shear stiffness [k_s(GPa·m^{-1})]	Cohesive strength [σ(MPa)]	Tensile strength [σ(kPa)]
Prestressed cable	130	195.0	0.2	450.0	12.45	12.45	5.20	620
Prestressed bolt	110	200.0	0.2	80.0	8.75	8.75	4.00	500
Grouting bolt	110	200.0	0.2	–	8.75	8.75	4.00	500
Steel pipe pile	127	210.0	0.2	–	24.5	24.5	7.20	–

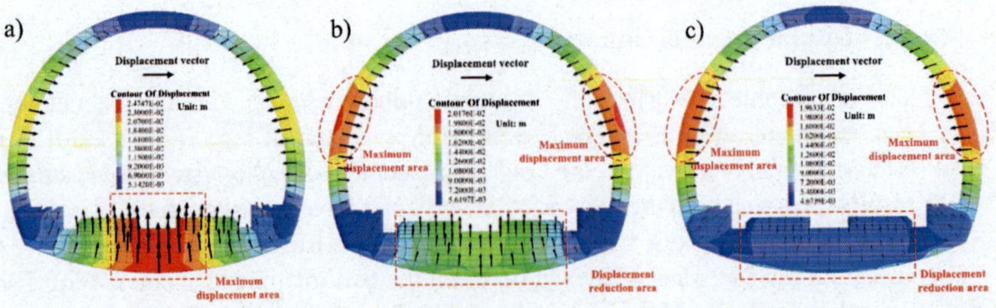

Fig. 14. Distribution of structure displacement with treatment. (a) Before tunnel renovation. (b) After tunnel renovation of Scheme 1. (c) After tunnel renovation of Scheme 2.

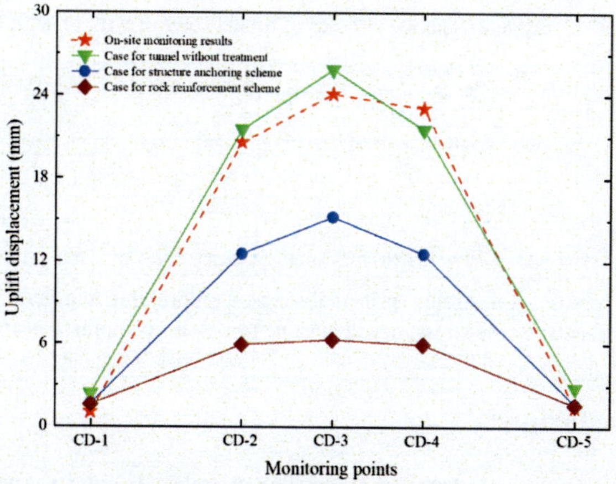

Fig. 15. Curve of uplift displacement of measuring points at bottom after tunnel renovation.

5.3 Structural Stress of Tunnel Support Structure

As an important evaluation index, the structural stress of tunnel support directly reflects the safety of tunnel structure. Table 8 shows the results of tunnel support stress with or without treatment measures at typical locations. According to the strength of support concrete (described in Sect. 4.2), those stress values exceeding the limit are bold, which means that the concrete at such locations was cracked. Such results show that after tunnel renovation, structural stresses at all measuring locations decrease significantly, and there is no cracking risk or crushing risk in bottom structure. Only from the perspective of structural safety, the above schemes can control the structural stress in a reasonable range to ensure the safety of tunnel support. However, the σ_1 of bottom structure with tunnel renovation of Scheme 2 is smaller than that of Scheme 1, which means that the Scheme 2 better ensure the safety of the tunnel bottom.

Generally, the Scheme 2 with a smaller uplift displacement at bottom and a safe structure stress is more suitable for the renovation of Yunding Tunnel. Such results also show that in view of the invert heave in layered rock masses, it is more useful to strengthen the mechanical parameters of rock mass and structural plane and weaken the layered characteristics of rock structure by grouting at tunnel bottom than to anchor the bottom structure with surrounding rock by prestressed cables.

Table 9. Computed concrete stresses of tunnel support with or without tunnel renovation.

Case	Lining of vault (C25)		Lining of arch foot (C25)		Lining of side wall (C25)		Invert filling (C20)		Track slab (C40)	
	σ_1 (MPa)	σ_2 (MPa)	σ_1 (MPa)	σ_2 (MPa)	σ_1 (MPa)	σ_2 (MPa)	σ_1 (MPa)	σ_2 (MPa)	σ_1 (MPa)	σ_2 (MPa)
Without tunnel renovation	−1.65	−16.33	0.47	−8.95	−4.57	−15.53	**1.92**	−6.32	**2.76**	−5.62
With Scheme 1	−1.44	−14.42	0.21	−8.32	−4.34	−9.83	1.23	−5.11	0.93	−4.32
With Scheme 2	−1.43	−14.65	0.09	−8.11	−4.22	−9.62	0.82	−4.75	0.61	−4.05

Note: σ_1 represents maximum principal stress; σ_2 represents minimum principal stress; positive value represents tensile stress; negative value represents compressive stress.

6 Conclusions

This paper investigated the behavior of an operation railway tunnel located in horizontal layered rock masses. The results of field investigation and numerical analysis were discussed and compared to explore the stress and displacement response of the tunnel support. Different factors of tunnel geological condition, such as layer thickness and lateral pressure coefficient, and two treatment schemes were investigated. The major findings and conclusions were derived as follows:

1. The deformation characteristics of layered rock masses was the main cause for the instability of bottom structure in the operating railway tunnel. Unlike homogeneous rocks, warping deformation of layered rock mass caused the bottom structure to deform into the tunnel, and formed the invert heave disease. Especially in the thin-thick to medium-thick layered rock mass, the uplift displacement of bottom structure could be more obvious. Such displacement resulted in a remarkable increase of tensile stress on the surface of bottom structure. When the maximum principal stress exceeded the allowable tensile strength of concrete, structural cracks appeared.
2. The influence of lateral pressure coefficient (K_O) on the stability of bottom structure was not simply unfavorable. When the K_O was low, the behavior of bottom structure was mainly affected by the warping deformation of layered rock mass below the invert, which lead to the large uplift displacement of bottom structure and the cracking risk of invert filling surface. With the increase of K_O, tunnel support was also subjected to the effect of horizontal compression deformation, which caused the uplift displacement to decrease at first and then increase, while the risk region of cracking shifts to a deeper position. When the K_O exceeded 3.0, not only the concrete of the vault and side wall might be crushed, but also the separation between the track slab and the invert filling may existed.
3. The treatment scheme of anchoring the bottom structure and surrounding rock restrained the uplift displacement of bottom structure to some extent, but did not change the deformation characteristic of layered rock mass. The rock reinforcement scheme by grouting at the bottom could strengthen the mechanical parameters of surrounding rock and structural plane and weaken the layered characteristics of rock masses. Obviously, the latter scheme better inhibits the invert heave and ensure

the safety of bottom structure, proving that the rock reinforcement scheme is more suitable for the tunnel renovation in horizontal layered rock masses.

Acknowledgments. This study was supported by the National Natural Science Foundation of China (Grant 51878669 and U1934211). The authors also appreciate the support of the Fundamental Research Funds for the Central Universities of Central South University (Grant 2019zzts291).

References

Adhikary, D.P., Dyskin, A.V.: Modelling the deformation of underground excavations in layered rock masses. Int. J. Rock Mech. Min. Sci. **34**, 3–4 (1997)

Ding, X., Niu, X., Pei, Q., Huang, S., Zhang, Y., Zhang, C.: Stability of large underground caverns excavated in layered rock masses with steep dip angles: a case study. Bull. Eng. Geol. Env. **78**(7), 5101–5133 (2019). https://doi.org/10.1007/s10064-018-01440-8

Do, N.A., Dias, D., Dinh, V.D., et al.: Behavior of noncircular tunnels excavated in stratified rock masses - case of underground coal mines. J. Rock Mech. Geotech. Eng. **11**, 99–110 (2019)

Fortsakis, P., Nikas, K., Marinos, V., et al.: Anisotropic behaviour of stratified rock masses in tunnelling. Eng. Geol. **141–142**, 74–83 (2012)

Itasca Consulting Group Inc.: FLAC3D User's Guide, 5th edn. Itasca Consulting Group Inc., Minneapolis (2012). www.itascacg.com

JGJ/T 384-2016: Chinese Technical Specification for Testing Concrete Strength with Drilled Core Method. China Architecture & Building Press, Beijing (2016)

Jia, C.J.: Study on the deformation detection and treatment of the bottom structure of Zhaojiayan Tunnel. Master's thesis. Jilin University (2018)

Jia, P., Tang, C.A.: Numerical study on failure mechanism of tunnel in jointed rock mass. Tunn. Undergr. Space Technol. **23**, 500–507 (2008)

Li, C.L.: Study on mechanism and deformation controlling technology of monoclinic interbed soft rock tunnel. Master's thesis. Shijiazhuang Tiedao University (2017)

Li, P.F., Liu, H.C., Zhao, Y., et al.: A bottom-to-up drainage and water pressure reduction system for railway tunnels. Tunn. Undergr. Space Technol. **81**, 296–305 (2018)

Li, A., Liu, Y., Dai, F., et al.: Continuum analysis of the structurally controlled displacements for large-scale underground caverns in bedded rock masses. Tunn. Undergr. Space Technol. **97**, 103288 (2020)

Małkowski, P.: The impact of the physical model selection and rock mass stratification on the results of numerical calculations of the state of rock mass deformation around the roadways. Tunn. Undergr. Space Technol. **50**, 365–375 (2015)

Moussaei, N., Sharifzadeh, M., Sahriar K., et al.: A new classification of failure mechanisms at tunnels in stratified rock masses through physical and numerical modeling. Tunn. Undergr. Space Technol. **91**, 103017 (2019)

Möller, S.C.: Tunnel induced settlements and structural forces in linings. Univ. Stuttgart, Institute of Geotechnik, Stuttgart (2006)

Qian, W.P., Qi, T.Y., Zhao, Y.J.: Deformation characteristics and safety assessment of a high-speed railway induced by undercutting metro tunnel excavation. J. Rock Mech. Geotech. Eng. **11**, 88–98 (2019)

Su, L.: Horizontal stratum stability analysis and construction measures of Maotianshan Tunnel. Railw. Constr. Technol. **2010**(3), 49–52 (2010)

Tang, H., Yong, R., Ez Eldin, M.A.M.: Stability analysis of stratified rock slopes with spatially variable strength parameters: the case of Qianjiangping landslide. Bull. Eng. Geol. Env. **76**(3), 839–853 (2016). https://doi.org/10.1007/s10064-016-0876-4

TB10003-2016/J449-2016: Code for Design of Railway Tunnel. China Railway Publishing House, Beijing (2016)

TB10012-2019/J124-2019: Code for Geology Investigation of Railway Engineering. China Railway Publishing House, Beijing (2019)

Wang, L.C., Xiao, X.W., Lin, H.: Analysis on Causes for and renovation of floor structure of a high-speed railway tunnel located in slightly-dipping interbedded rock mass. Tunn. Constr. **34**(9), 823–836 (2014)

Xu, Z.Q.: Study on influencing factors and countermeasure of tunnel bottom uplift in stratified rock mass with high ground stress. Master's thesis. Central South University (2019)

Yang, X.L., Zhou, T., Li, W.T.: Reliability analysis of tunnel roof in layered Hoek-Brown rock masses. Comput. Geotech. **104**, 302–309 (2018)

Zhang, Y.J., Ding, H.J.: Large-deformation mechanism and gountermeasures of soft rock in Gaopo Tunnel on Chengdu-Guiyang high-speed railway. Sci. Technol. Eng. **20**(1), 327–334 (2020)

Zhou, Y.Y., Xu, D.P., Gu, G.K., et al.: The failure mechanism and construction practice of large underground caverns in steeply dipping layered rock masses. Eng. Geol. **250**, 45–64 (2019)

A Semi-analytical Model for a Compaction-Grouted Soil Nail with Grout Bulb

Yu Li, Xin-yu Ye(✉), Rui Peng, and Sheng Zhang

School of Civil Engineering, Central South University, Changsha, Hunan, China
{Yuli0523,yexinyu113}@csu.edu.cn

Abstract. The densification of surrounding soil from compaction grouting for the compaction-grouted soil nail is a significant merit to improve the pullout force. However, few studies was conducted to quantify the contribution of the compaction grouting to the pullout force, thus, a semi-analytical model was proposed. First, a hyperbolic model considers the soil-grout bulb (soil nail) interface behavior was established, and an energy transfer analysis model of a soil nail with grout bulb was proposed by combining the energy method and the hyperbolic model. Second, the semi-analytical model was verified based on the result of physical model test. Third, parametric study was conducted to evaluate the influence of the key parameter of the hyperbolic model on the pullout force with grouting process and those without.

1 Introduction

Soil nailing is one of the most effective methods in slope stabilization, foundation pit supporting and retaining wall (Schlosser 1982), and the combination with grouting, which causes the complicated phenomenon of compaction and fracturing, is the most widely used soil nail type (Wang et al. 2016). However, to achieve the ideal compaction grouting, the geotextile or membrane is normally adopted at the injection tube. In this case, the grouting effect can be better controlled to prevent the penetration and splitting of the slurry in practice (Wang et al. 2009). To investigate the influence of the grouting effect, Au et al. (2004) observed that the soil consolidation by grouting is due to the greater stress concentration in the surrounding soil by using an improved consolidation instrument. Yin et al. (2012) presented a method considering grouting pressure and soil densification to calculate the maximum shear stress of soil-nail interface. Wang et al. (2013) investigated the influence of the compaction effect caused by grouting pressure on the void ratio or consolidation of the surrounding soil by designing the apparatus with a function of dynamic compaction.

Many studies have been done to understand the influence of compaction effect on the surrounding soil. However, most of then failed to further investigate the improvement of compaction-grouted soil nail/pile bearing capacity derived from compaction effect. Yin et al. (2009) performed field tests for conventional soil nail and concluded that the compaction effect on the surrounding soil is the main reason for the enhancement

J. J. Yang et al. (Eds.): GeoChina 2021, SUCI, pp. 57–70, 2021.
https://doi.org/10.1007/978-3-030-79672-3_4

of the pullout force of a soil nail. Zhang et al. (2014) stated that the diameter of the slurry bag (formed by compaction grouting) can improve performance of the so-called bag soil nail and the length of cured grout body can improve the pullout force of soil nail. Wang et al. (2017) proposed a new type of soil nail which has a grout bulb after grouting at the injection hole with a latex balloon, and found that this new soil nail with grout bulb has better performance than the traditional soil nail based on the results in laboratory. For this new soil nail, Ye et al. (2019a) also conducted physical model test to study the influence of the grouting pressure on the pullout force. In addition, many numerical simulations are also carried out to investigate the influence of shape of bubble on compaction effect. The relationship between the shape of bolt and the shear strength of bolt-cement interface was conducted by a two-dimensional discontinuous deformation analysis (2D-DDA) program (Nie et al. 2019; Yokota et al. 2019). By using a verified numerical model, the pullout force of the soil nail was studied to investigate the influence of soil dilation, which is influenced by compaction grouting on surrounding soil (Su et al. 2010). Xue et al. (2013) conducted the interface between soil nail and surrounding soil by using the modified Goodman element. Based on the results of numerical simulation and laboratory test, Ye et al. (2017, 2019b) investigated the pullout behavior of the soil nail with single, and multiple grout bulbs. However, the compaction effect of the previous studies regarding the conventional soil nail has little influence on the improvement of a soil nail, and the action from grouting on surrounding soil is not an ideal compaction process, which is different from the compaction-grouted soil nail. On the other hand, the studies on compaction-grouted soil nail failed to establish a relationship between the enhancement of pullout force and compaction effect. Therefore, a study which can quantify the contribution of compaction effect (i.e., densification from grouting) to the pullout force is needed.

In this study, a semi-analytical model was developed to calculate the pullout force of compaction-grouted soil nail, which can well describe the effect of compaction effect on pullout force. Firstly, based on the tests with and without compaction effect, the enhancement of compaction effect on pullout force is analyzed. Secondly, the nonlinear relationship between grouting pressure and compaction effect is described by an exponential model. Thirdly, the calculation model of pullout force is established based on energy equilibrium theory, and the proposed model is verified by the test results. Finally, the influence of compaction effect on pullout force of the soil nail is further discussed.

2 Energy Equilibrium Equation of the Soil Nail with Grout Bulb Considering the Compaction Effect

2.1 Establish of Relationship Considering Compaction Effect

Figure 1 compares the profiles of the pullout force with displacement for the tests with and without compaction effect under a grout pressure of 500 kPa, 600 kPa, respectively. The test data are available in the literatures (Wang et al. 2017; Ye et al. 2017), and the detailed summaries of the experiment scheme for these tests can be found in Ye et al. (2017). The curves of pullout force versus displacement with compaction effect are apparently higher than those without compaction effect under the same grouting pressure. The difference of the pullout force of the soil nail with and that without the

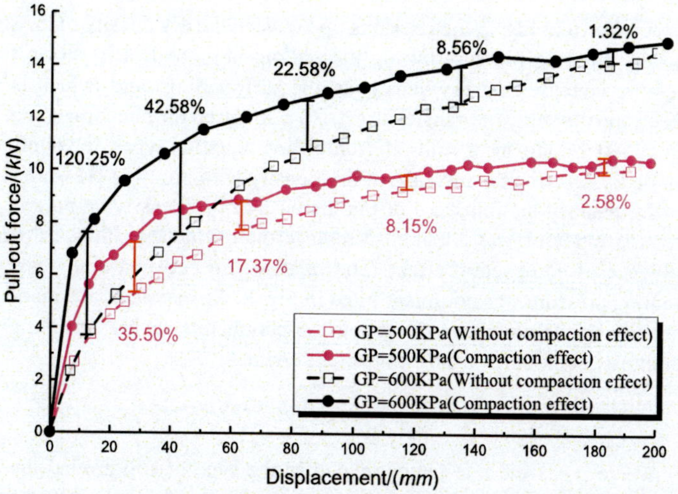

Fig. 1. Comparison of pullout force between compression effect and non-compression effect

compaction effect is shown mainly in the displacement range of 0–40 mm, and the difference rate ((compaction effect-no compaction effect)/no compaction effect) of the pullout force, under the grouting pressure of 500 kPa, 600 kPa, are 35.50% and 120.25%, respectively. In the displacement range from 40 to 140 mm, the mobilized pullout forces were reduced to 8.15% and 8.56%, respectively. The difference between the value of the pullout force with compaction effect and that without is extremely small, only 2.58% and 1.32%. This is because the dry density of sand surrounding the grout bulb, when the grouting pressure increasing from 500 kPa to 600 kPa, is caused by compaction effect. Therefore, the dry density of the soil around the grout bulb increases, which increases the pullout force of the compaction-grouted soil nail.

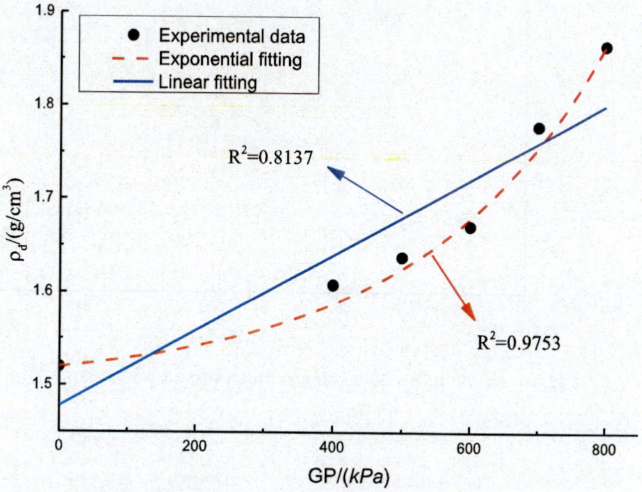

Fig. 2. Relationship between grouting pressure and dry density

As shown in Fig. 2, the relationship between the dry density of the surrounding sand and the grouting pressure during the pullout is plotted. It is shown that the grouting pressure increases the dry density of the surrounding sand, which is the compaction effect, and the study of Wang et al. (2016) also raised the phenomenon of the compaction effect. After comparing with different fitting equation types, the nonlinear relationship can better describe the increase of dry density with grouting pressure.

To quantify the influence of the increase of dry density on pullout force, the compression modulus was adopted to replace the dry density, thus, to better describe the influence of compaction effect. The relationship between compression modulus and grouting pressure was nonlinear fitted in Fig. 3. An exponential equation (Eq. 1) is used considering the nonlinear change of the parameters such as grouting pressure, initial compression modulus from compaction effect.

$$E_s = E_0 e^{(Gp - Gp_0)/(\mu_1 * Gp_0)} \tag{1}$$

where μ_1 is the rate factor related to the type of soil around the soil nail, as the cohesion is zero for sand, the value of sand (about 0.8) is generally smaller than that of clay (about 20% smaller) (Abchir et al. 2016). In this study, the value of 0.7 was selected; GP is the grouting pressure; E_0 is the initial compression modulus corresponding to the compression modulus without compaction effect and the case that the grouting pressure is GP_0. In this study, 320 kPa was adopted according to the test. Because the grout slurry is difficult to be effectively injected into the membrane due to the loss of grouting pressure during grouting and the restriction of surrounding soil. The results of several model tests also proved that grout slurry was hard injected when the grouting pressure is less than 400 kPa.

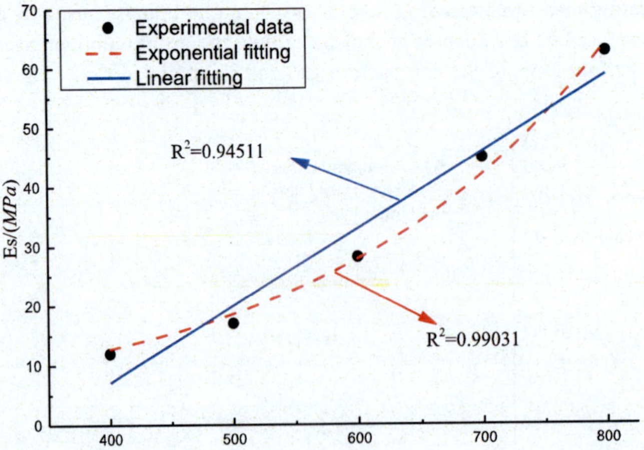

Fig. 3. Relationship between grouting pressure and compression modulus

Based on Carrubba et al. (1997), the initial slope ($1/a_1$) expression of the hyperbolic model of the surface of the grout bulb and soil nailing rod is obtained.

$$1/a_1 = G_1/R\ln(R_0/R) \tag{2}$$

The initial slope ($1/a_1$) expression of the hyperbolic model of the curved surface of the grout bulb is obtained:

$$1/a_1 = E_1/R\ln(R_0/R) \tag{3}$$

Where R is the radius of the soil nail; R_0 is the influence range of the radius of the soil nail; E_1 and G_1 are compression modulus and shear modulus, respectively. The value of $\ln(R_0/R)$ is 3–5 (Baguelin et al. 1979), and the value is 3 in this paper.

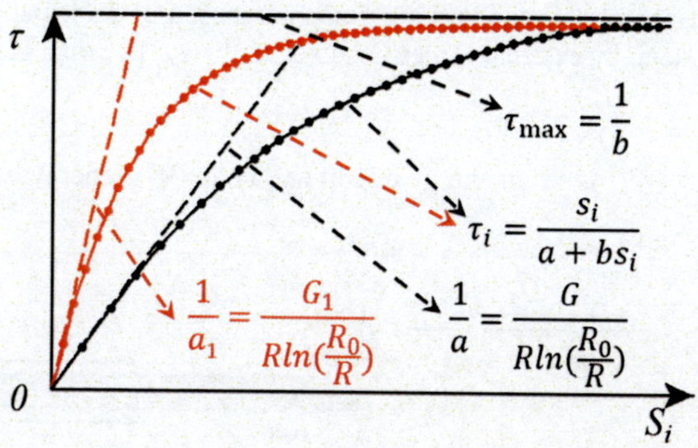

Fig. 4. Hyperbolic model

Through the analysis of the mechanism of compaction effect and the establishment of compaction effect model, the relationship between the side resistance of compaction-grouted soil nail with grout bulb and the relative displacement of soil can be obtained (Fig. 4).The difference from the no compaction effect is that the initial slope ($1/a_1$) becomes larger, and the function expression is:

$$\tau_i = \frac{s_i}{a_1 + bs_i} \tag{4}$$

In Eq. (4), a_1 and b are the parameters of the hyperbolic curve model, the reciprocal of a_1 and b represent the initial slope of the hyperbolic curve and the ultimate shear strength of soil, respectively. S_i is the pull-our displacement at any point, and τ_i is the shear strength of the pullout displacement S_i.

The expressions of the parameter ($1/b$) of the surface of the grout bulb and soil nail rod are as follows:

$$1/b = \tau_{max} \tag{5}$$

The expression of the parameter $(1/b)$ of the curved surface of the grout bulb is as follows:

$$1/b = \sigma_{ult} \tag{6}$$

where $\tau_{\max}$ and σ_{ult} are the ultimate shear strength and the ultimate compression strength around the soil nail, respectively.

2.2 Analysis of Soil Nail with a Grout Bulb

Because of the existence of the grout bulb in the soil nail, it becomes complicated to consider the stress state of the soil nail with variable cross-section by using the previous solution method. Moreover, it is inefficient to solve the force balance equation and the displacement coordination equation of the rod element simultaneously. Therefore, the energy method is used to study the whole soil nailing-soil system (Fig. 5), then the total energy change of soil nail-soil system is composed of deformation energy of soil nail and work done by external forces (Zhao et al. 2008). The total balance equation is:

$$\Pi = U + W \tag{7}$$

In Eq. (4), U is strain energy of soil nail body; W is energy by external force.

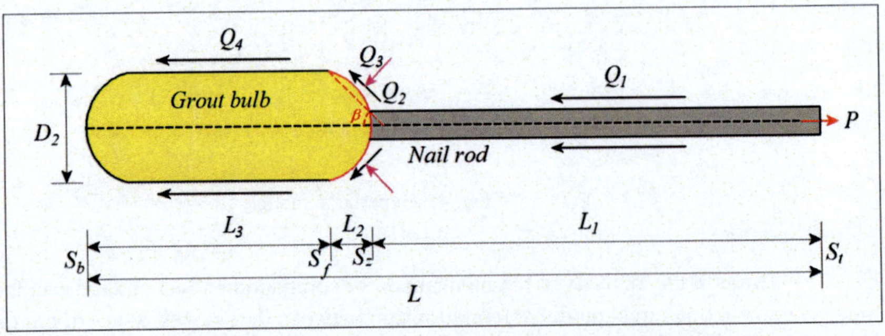

Fig. 5. Schematic diagram of the soil nail with grout bulb

As shown in Fig. 5, L is the soil nailing depth; P is the pullout force of the soil nail; D_1 and L_1 are the diameter and length of the soil nail rod, respectively; D_2 is the diameter of the grout bulb; the length of the curved surface of the grout bulb is $L_2 - L_1$; the length of the plane of the grout bulb is $L_3 - L_2$. The following assumptions are made to simplify the derivation process: (1) soil nail and soil are isotropic and homogeneous materials; (2) the curved surface of the grout bulb is assumed to be linear; (3) the ultimate compression strength of soil does not change with the increase of grouting pressure.

The section diameter of the soil nail varies due to the grout bulb. Therefore, the soil nail with grout bulb is divided into soil nail rod, the curved surface, and the plane of the grout bulb. The strain energy of the whole soil nail is as follows:

$$U = \alpha_1 \int_0^{L_1} \left(\frac{d\delta}{dx}\right)^2 dx + \alpha_2 \int_{L_1}^{L_2} \left(\frac{d\delta}{dx}\right)^2 dx + \alpha_3 \int_{L_2}^{L_3} \left(\frac{d\delta}{dx}\right)^2 dx \tag{8}$$

In Eq. (8), $\alpha_1 = \pi E_1 D_1^2/8$, $\alpha_2 = \pi E_2(4(L_2^3 - L_3^3)\tan\beta + 3dD_1^2(L_2 - L_1) + 6D_1(L_2^2 - L_1^2)\tan\beta)/48$, $\alpha_3 = \pi E_3 D_2^2/8$, E_1, E_2, E_3 are the elastic modulus of soil nail rod, the curved surface and the plane of the grout bulb, respectively.

The forces on the curved surface of the soil nail with grout bulb are analyzed according to the semi-infinite space theory of the literature (Cao et al. 2013). As shown in Fig. 6, the soil nail is displaced by the load P, and the force on the slope is decomposed into the normal and shear stresses (Q_3, Q_2) at the curved surface of the grout bulb for static equilibrium of the soil nail. Therefore, The energy W by the external force can be obtained as:

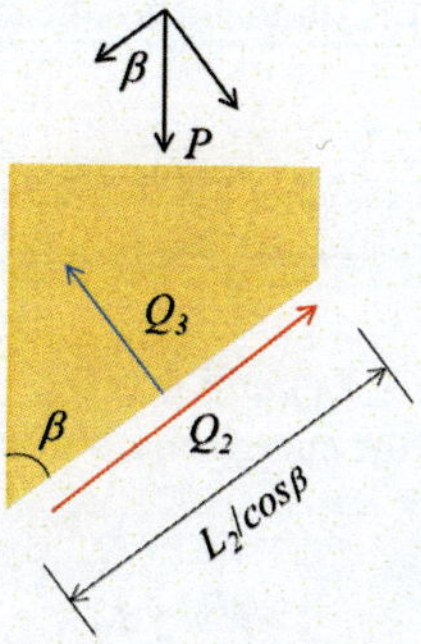

Fig. 6. Free body diagram of the curved surface of grout bulb

$$W = W_f + W_F + W_P \tag{9}$$

$$W_f = -\iint \tau(x)(\delta + s_b)ds \tag{10}$$

$$W_F = -\iint Q_3(s_b + \delta(L_3 - L_1)/L)\sin\beta ds \tag{11}$$

$$W_P = Ps_t \tag{12}$$

Where W_f, W_F and W_P are the the energy by the side resistance, the normal stresses Q_3 and the load P, respectively.

The whole energy equilibrium equation can be obtained through simultaneous Eqs. (7)–(12).

$$Ps_t = \alpha_1\int_0^{L_1}\left(\frac{d\delta}{dx}\right)^2 dx + \alpha_2\int_{L_1}^{L_2}\left(\frac{d\delta}{dx}\right)^2 dx + \alpha_3\int_{L_2}^{L_3}\left(\frac{d\delta}{dx}\right)^2 dx + \iint \tau(x)(\delta + s_b)ds - \iint Q_3\left(s_b + \frac{\delta(L_3 - L_1)}{L}\right)\sin\beta ds \tag{13}$$

The energy equilibrium equation of soil nail rod ($0 \le x \le L_1$) is obtained according to Eq. (13):

$$P_i s_i - P_{i+1}s_{i+1} = \alpha_1\int_0^{L_1}\left(\frac{d\delta}{dx}\right)^2 dx + \iint \tau(x)\left(\frac{d\delta}{2} + s_{i+1}\right)ds \tag{14}$$

The Eq. (15) can be obtained by the method of undetermined coefficients in Eq. (14):

$$s_{i+1} = s_i - \frac{2hP_i - \pi D_1 h^2 \int_i^{i+1} \tau(x)dx}{2\alpha_1} \tag{15}$$

The energy equilibrium equation of the curved surface of the grout bulb ($L_1 \le x \le L_2$) is obtained according to Eq. (13):

$$P_i s_i - P_{i+1} s_{i+1} = \alpha_2 \int_{L_1}^{L_2} \left(\frac{d\delta}{dx}\right)^2 dx + \iint \tau(x)\left(\frac{d\delta}{2} + s_{i+1}\right)\cos\beta ds + \iint Q_3\left(s_b + \frac{\delta(l_3 - l_1)}{L}\right)\sin\beta ds \tag{16}$$

Similarly, the Eq. (17) is obtained by the method of undetermined coefficients in Eq. (16):

$$s_{i+1} = s_i - \frac{2hp_i - \omega h^2\left(\int_i^{i+1} \tau(x)dx\cos\beta + \int_i^{i+1} Q_3(x)dx\sin\beta\right)}{2a_2} \tag{17}$$

In Eq. (17): $\omega = \pi(nh - L_1)tan\beta + \pi D_1$.

According to Eq. (13), the energy equilibrium equation of the plane of the grout bulb ($L_2 \le x \le L_3$) is obtained:

$$P_i s_i - P_{i+1} s_{i+1} = \alpha_3 \int_{L_2}^{L_3} \left(\frac{d\delta}{dx}\right)^2 dx + \iint \tau(x)\left(\frac{d\delta}{2} + s_{i+1}\right)ds \tag{18}$$

Similarly, by means of undetermined coefficients in Eq. (18), the Eq. (19) can be obtained:

$$s_{i+1} = s_i - \frac{2hP_t - \pi d_2 h^2 \int_i^{i+1} \tau(x)dx}{2a_3} \tag{19}$$

Therefore, the expression of displacement and pullout resistance of soil nail with grout bulb is obtained:

$$s_{i+1} = \begin{cases} s_i - \dfrac{2hP_i - \pi D_1 h^2 \int_i^{i+1}\tau(x)dx}{2\alpha_1} & 0 \le x \le L_1 \\ s_i - \dfrac{2hP_i - \omega h^2\left(\int_i^{i+1}\tau(x)dx\cos\beta + \int_i^{i+1}Q_3(x)dx\sin\beta\right)}{2\alpha_2} & L_1 \le x \le L_2 \\ s_i - \dfrac{2hP_i - \pi d_2 h^2 \int_i^{i+1}\tau(x)dx}{2\alpha_3} & L_2 \le x \le L_3 \end{cases} \tag{20}$$

Because each segment of the soil nail is continuous, the relationship between side resistance and displacement (1–6) of each segment is substituted into the Eq. (20) to obtain the equilibrium equation of the pullout displacement and the pullout force of the soil nail with the grout bulb. The curves of pullout force and pullout displacement of the soil nail can be obtained by the boundary conditions ($P_{n+1} = 0$ and $P_1 = P$).

3 Laboratory Test Verification

3.1 Test System

A series of large-scale model tests for a compaction-grouted soil nail was carried out at the University of Newcastle. The laboratory test system, which consists of five parts (test model box device, overburden pressure loading device, pullout loading device, grouting device, data acquisition device), was shown in Fig. 7. As shown, the overburden pressure loading device consists of the compressed air source, a rigid PVC sheet, linear variable differential transformers (LVDTs), rubber bag and regulator. pullout loading device consists of hydraulic jack and linear variable differential transformer (LVDT). The grouting device consists of grouting pipe, slurry cylinder, scale, compressed air source and valve.

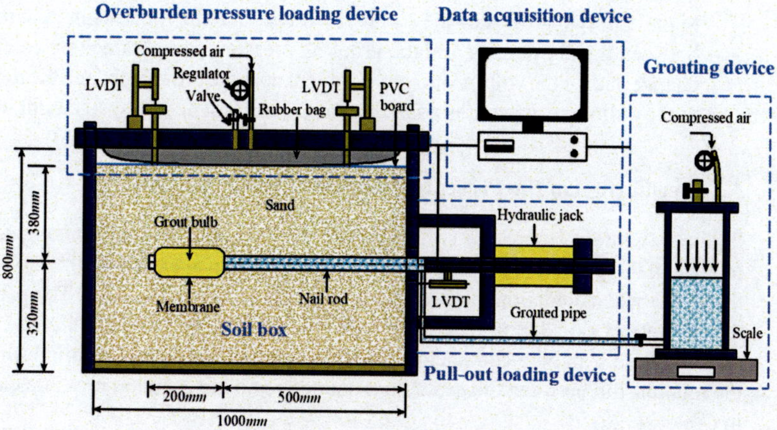

Fig. 7. Schematic diagram of the model test system

The soil box, with a length of 1000 mm, width of 60 mm and height of 800 mm, was used to prepare the sand sample, with the sand from Stockton Beach, Newcastle (the grain size distribution and mechanical parameters are available in Ye et al. (2017, 2019b)). The compaction-grouted soil nail was installed during the preparation of soil sample. The cement slurry with a water/cement ratio of 0.5 was selected and injected using a cylinder connected with compressed air.

3.2 Test Procedure

The main test steps are as follows:

(1) Sample filling: The prepared soil samples were layered and compacted, which the thickness of those soil samples for each soil layer was controlled to about 40 mm. Meanwhile, the soil nail and the soil physical parameter sensors were respectively

installed in the process of sample filling. In the control group without compaction effect, it should be noted that the volume of the grout bulb was equal with or without compaction effect.

(2) Overburden pressure. The overburden pressure of 100 kPa in model test, through the overburden pressure loading device on the upper part of the model box, was applied to simulate the overburden pressure at a certain depth of the compaction-grouted soil nail with grout bulb.
(3) Grouting: The grouting pressure (GP) was 400 kPa, 500 kPa, 600 kPa, 700 kPa and 800 kPa, respectively, when the installed soil nail was injected with the adjusted compressed air. The no compaction effect was compared with the compaction effect by controlling the volume of the slurry (the volume of the grout bulb).
(4) Pull out: After 7 days of cement slurry curing, two groups of soil nail with grout bulb tests (compaction effect and no compaction effect) were carried out by hydraulic jack at a rate of 1 mm/min. At the rate of 1 mm/min, the test drawing load was used to pull out two groups of the soil nail with grout bulb. Meanwhile, the relationship between the dry density and the grouting pressure was obtained by calculating the change data of the volume moisture content near the grout bulb, and the development of the pullout resistance as a function of the soil nail head displacement was plotted.

3.3 Verification of the Presented Method

Figure 8 (without compaction effect) and Fig. 9 (compaction effect) compares the difference of the pullout force plotted as a function of displacement for the model tests and the theoretical calculation method under different grouting pressures. The parameters of the theoretical calculation method are shown in the Table 1, where the no compaction effect is Case 1 and the compaction effect is Case 2. In addition, detailed summaries of the relationship between the grout bulb diameter and the grouting pressure can be found in (Ye et al. 2017).

Table 1. The analysis parameters of validation case

Case	Shape features						Soil characteristics			
	L (m)	D_1(m)	L_1(m)	E (kN/m)^2)	D_2(m)	β(deg)	$1/b$ (kPa)	E_0 (MPa)	G_0 (MPa)	μ_1
1	0.7	0.035	0.472	1.8e6	/	20	20	34	5	/
2	0.7	0.035	0.472	1.8e6	/	20	50	34	7	0.7

As shown in Fig. 8 and Fig. 9, the calculated pullout force by the analytical model matches that of the laboratory test well. However, some differences are exhibited between the calculated and experimental pullout forces at the displacement that smaller than 40 mm. The main reason is that the shape of the grout bulb formed by grouting is irregular (Ye et al. 2020), so the side resistance between the grout bulb and surrounding soil is larger than predicated since the analytical model adopts the regular shape of the grout bulb.

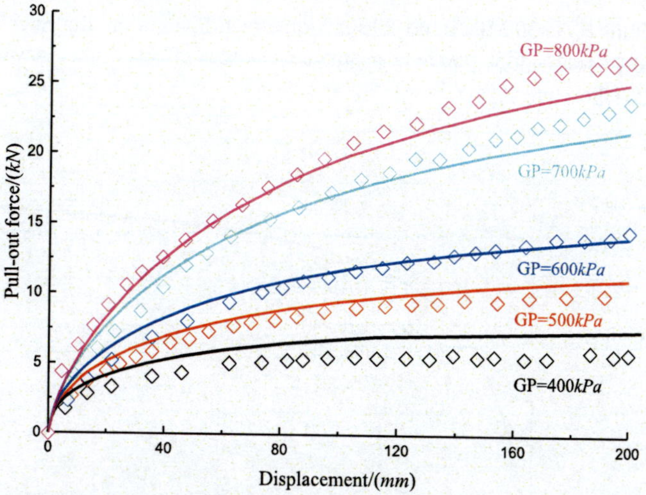

Fig. 8. The comparison of the laboratory model tests (without compaction effect) and the analytical method. The solid lines are the predicted results by the analytical model, and the symbols indicate the measured results.

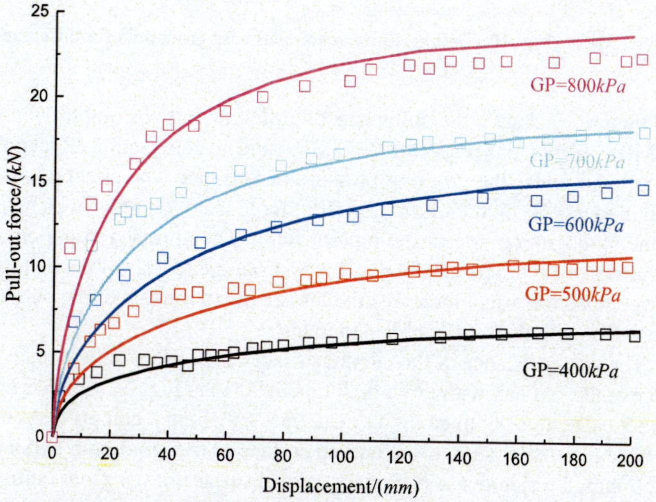

Fig. 9. The comparison of the laboratory model tests (compaction effect) and the analytical method. The solid lines are the predicted results by the analytical model, and the symbols indicate the measured results.

3.4 Parameter Analysis

The pullout law of the soil nail with the grout bulb is discussed by changing the parameters of initial compression modulus. The values of the parameters are the same as those of Case 2, and the values of the no compaction effect are Case 1. The initial compression modulus E_0 is 30 MPa in no compaction effect. In addition, the calculation parameters of the presented method are re-evaluated when the effects of the relevant parameters are analyzed.

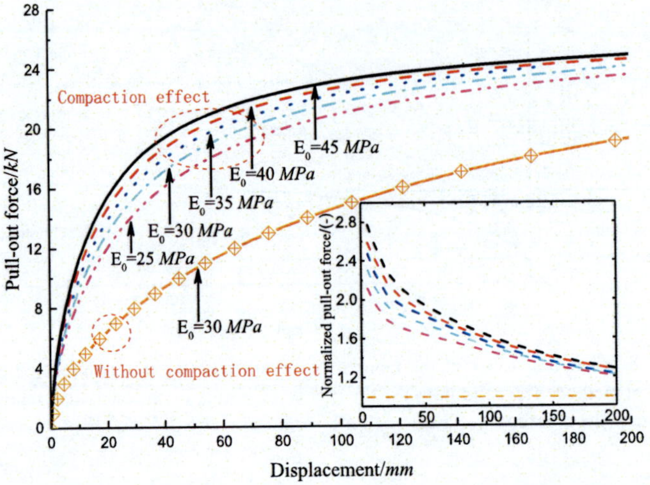

Fig. 10. Pullout force of soil nail with grout bulb for different E_0

Figure 10 shows the influence of different initial compression modulus (E_0 = 25 MPa, 30MPa, 35MPa, 40MPa, 45MPa) and no compaction effect on the pullout force of soil nail under the grouting pressure of 800 kPa. The effect of the initial compression modulus E_0 on the ultimate pullout force is not obvious, which is mainly reflected in the mobilization rate of the pullout force. The normalized pullout force shows that the effect of E_0 progressively decreases (from 2.8 to 1.2) as the displacement increase. The compaction effect increases the rate mobilization rate of the pullout force, which is consistent with the results obtained in tests.

The side resistance of the curved surface of the grout bulb is main source of pullout force of the soil nail with grout bulb (about 90%) in Fig. 11, which is consistent with the numerical results obtained in (Ye et al. 2017). the initial compression modulus increases the side resistance ratio of the curved surface of the grout bulb. After the displacement of 50 mm, the side resistance of the curved surface of the grout bulb remains basically constant (90%). By compacting the soil, the initial compression modulus is increased, and the mobilization rate of the pullout resistance of the soil nail is increased. In practice, the pullout displacement is very small, so the pullout force of the soil nail with grout bulb is improved.

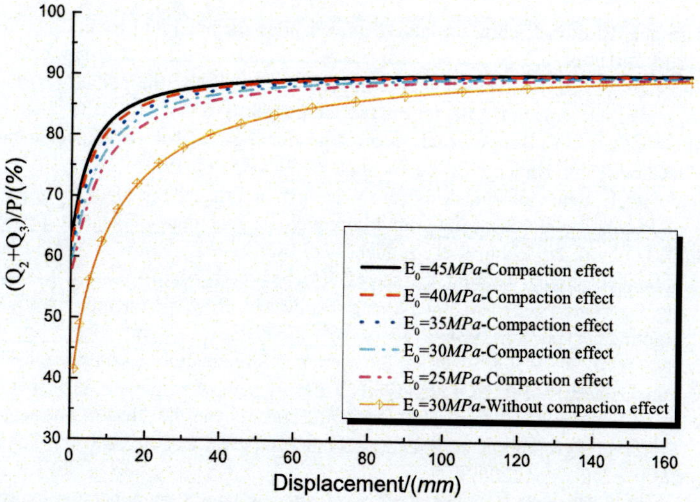

Fig. 11. Side resistance of the curved surface of the grout bulb for different E_0

4 Conclusions

An exponential relationship was used to describe the variation of compaction effect under different grouting pressure, based on which the energy equilibrium equation considers compaction effect is derived by energy principle. Then, an analytical model for the pullout force with displacement was proposed. In addition, the analytical model shows a good performance in calculating the pullout force with and without compaction effect (i.e., grouting process) based on the verification of the results of laboratory test. Furthermore, the parametric study shows that the initial compression modulus influences the initialized rate of the pullout force significantly, but the influence on the ultimate pullout force (corresponding to the displacement of 200 mm) is not obvious. The side resistance of the curved surface of the grout bulb contributes the most to the pullout force throughout the pull-out process. In this paper, the study was conducted in sand, the initial compression modulus parameter of other types of soil can be done in the future.

References

Abchir, Z., Burlon, S., Frank, R., et al.: t–z curves for piles from pressuremeter test results. Géotechnique **66**(2), 137–148 (2016)

Au, S.K.A., Soga, K., Bolton, M.D., Jafari, M.R.: Laboratory investigation of multiple grout injections into clay. Géotechnique **54**(2), 81–90 (2004)

Baguelin, F., Frank, J.: Theoretical studies of piles using the finite element method. In: Institution of Civil Engineers Conference on Numerical Methods in Offshore Piling, London, pp. 83–91 (1979)

Cao, C., Nemcik, J., Aziz, N., et al.: Analytical study of steel bolt profile and its influence on bolt load transfer. Int. J. Rock Mech. Min. Sci. **60**, 188–195 (2013)

Carrubba, P.: Skin friction on large-diameter piles socketed into rock. Can. Geotech. J. **34**(2), 230–240 (1997)

Hong, C., Yin, J., Pei, H., Zhou, W.: Experimental study on the pullout resistance of pressure-grouted soil nails in the field. Can. Geotech. J. **50**(7), 693–704 (2013)

Nie, W., Zhao, Z.Y., Guo, W., et al.: Bond-slip modeling of a CMC rockbolt element using 2D-DDA method. Tunn. Undergr. Space Technol. **85**, 340–353 (2019)

Schlosser, F.: Behaviour and design of soil nailing. In: Proceedings of the International Symposium on Recent Development in Ground Improvement Techniques, Bangkok, pp. 399–413 (1982)

Su, L.J., Yin, R.H., Zhou, R.H.: Influences of overburden pressure and soil dilation on soil nail pullout resistance. Comput. Geotech. **37**(4), 555–564 (2010)

Wang, Q., Wang, S., Sloan, S.W., Sheng, D., Pakzad, R.: Experimental investigation of pressure grouting in sand. Soils Found. **56**(2), 161–173 (2016)

Wang, Q., Ye, X., Wang, S., et al.: Experimental investigation of compaction-grouted soil nails. Can. Geotech. J. **54**(12), 1728–1738 (2017)

Wang, S.Y., Chan, D.H., Lam, K.C.: Experimental study of the effect of fines content on dynamic compaction grouting in completely decomposed granite of Hong Kong. Constr. Build. Mater. **23**(3), 1249–1264 (2009)

Wang, S.Y., Chan, D.H., Lam, K.C., et al.: A new laboratory apparatus for studying dynamic compaction grouting into granular soils. Soils Found. **53**(3), 462–468 (2013)

Xue, X., Yang, X., Liu, E.: Application of the modified Goodman model in soil nailing. Int. J. Geomech. **13**(1), 41–48 (2013)

Ye, X., Wang, S., Wang, Q., et al.: Numerical and experimental studies of the mechanical behaviour for compaction grouted soil nails in sandy soil. Comput. Geotech. **90**, 202–214 (2017)

Ye, X., Wang, S., Wang, Q., et al.: The influence of the degree of saturation on compaction-grouted soil nails in sand. Acta Geotech. **14**(4), 1101–1111 (2019)

Ye, X., Wang, S., Xiao X., et al.: Numerical study for compaction-grouted soil nails with multiple grout bulbs. Int. J. Geomech. **19**(2), 04018193.1–04018193.11 (2019b)

Ye, X., Wang, S., Zhang, S., et al.: The compaction effect on the performance of a compaction-grouted soil nail in sand. Acta Geotechnica (5) (2020)

Yin, J., Hong, C., Zhou, W.: Simplified analytical method for calculating the maximum shear stress of nail-soil interface. Int. J. Geomech. **12**(3), 309–317 (2012)

Yin, J.H., Su, L.J., Cheung, R.W.M., et al.: The influence of grouting pressure on the pullout resistance of soil nails in compacted completely decomposed granite fill. Géotechnique **59**(2), 103–113 (2009)

Yokota, Y., Zhao, Z., Shang, J., et al.: Effect of bolt configuration on the interface behaviour between a rock bolt and bond material: a comprehensive DDA investigation. Comput. Geotech. **105**, 116–128 (2019)

Zhang, X., Wu, X., Yu, J., et al.: New slurry pressure type soil-nailing technology and its working mechanism. Chin. J. Geotech. Eng. **36**(S2), 227–232 (2014). (in Chinese)

Zhao, M., Liu, S.: Numerical simulation of negative skin friction on single pile in multiple layer deposits. Chin. J. Geotech. Eng. (03), 336–340 (2008). (in Chinese)

Prediction on the Penetration Resistance of Mono-Bucket Foundations in Silty Soil

Kanmin Shen[1], Tao Li[1], Kuanjun Wang[1](✉), Bin Wang[1], Jian Yu[2], and Zhigang Shan[1]

[1] Key Laboratory of Far-Shore Wind Power Technology of Zhejiang Province, POWERCHINA Huadong Engineering Corporation Limited, Hangzhou 311122, Zhejiang, China
{li_t2,wang_b6,shan_zg}@ecidi.com

[2] Department of Geotechnical Engineering, Key Laboratory of Geotechnical and Underground Engineering of Ministry of Education, Tongji University, Shanghai 200092, China
002yujian@tongji.edu.cn

Abstract. Hybrid mono-bucket foundations with ballast and skirt are employed to support offshore wind turbines (OWTs) in East China, with the advantages in capacity and efficient installation. The prediction of soil resistance during the penetration is critical in the design of mono-buckets, but few studies focused on the installation in silty soil. Two methods on predicting the penetration resistance are introduced based on soil parameters and in-situ CPT results, respectively. A modification on the friction angle of silty soil is made to account for the effect of the cohesion on the skirt-soil interface. A comparison is then carried out between the prediction results and prototype installation data to verify the methods and calibrate the coefficients in the equations. The back-analysis according to the field installation data shows that the accuracy of the first method is improved by the modification, and the second method based on the in-situ CPT data has a higher reliability with reasonable coefficients.

Keywords: Mono-bucket foundation · Offshore wind turbine · Penetration resistance · Suction installation · Cone penetration test · Silty soil

1 Introduction

Over the past 10 years, offshore wind power has been developing rapidly in China. Offshore wind turbines (OWTs) with larger capacity tend to be installed in deep water, leading to enormous overturning moment acting on the foundation (Byrne et al. 2003). Therefore, foundation design would encounter challenge in clayey and silty seabed with low soil strength and modulus. The rough weather condition in far-shore region further limit the application of traditional installation method as pile driving.

Due to the convenience in installation, skirted foundations with different geometry dimensions, including suction anchors and caissons, are increasingly used on offshore floating and fixed structures (Tjelta 2015; Andersen et al. 2005; Randolph and Gourvenec 2017). Normally the mono-bucket foundation would be inclined subjected to large lateral loads on the OWT (Nielsen et al. 2015). Hence, the jacket foundation with three or four

J. J. Yang et al. (Eds.): GeoChina 2021, SUCI, pp. 71–82, 2021.
https://doi.org/10.1007/978-3-030-79672-3_5

buckets are often used for large capacity turbines (Sturm 2017). However, they would increase the construction cost of the steel structure.

In view of above, a novel concept of hybrid mono-bucket foundation is proposed and adopted to support OWTs in China, especially in those regions with soft soils. As shown in Fig. 1(a), the mono-bucket foundation consists of a transition piece and a steel skirt. The transition piece is made of concrete and steel, and can contribute the self-weight of the mono-bucket foundation. The inner stiffeners separate the inner space into rooms and prevent the structure from bucking. The mono-bucket could resist the enormous overturning moment caused by wind load even in soft clay, with an outer diameter of over 30 m. The integrated installation process, as shown in Fig. 1(b), is a significant advantage of mono-bucket. The upper structure including rotor nacelle, blades and tower can be assembled together with the foundation at the harbour before shipment (Ding et al. 2015; Zhang et al. 2015). Till now, over 10 mono-bucket foundations have been installed in wind farms in East China, mainly in silty soil.

(a) Hybrid mono-bucket foundation (b) Mono-buckets during shipment

Fig. 1. Hybrid mono-bucket foundation during construction and shipment

The penetration mechanism of suction buckets has been well studied both in sand (Houlsby et al. 2005a; Tran et al. 2007; Senders et al. 2007, 2009) and in clay (Houlsby et al. 2005b; Guo et al. 2012). However, few studies have been conducted on the penetration resistance in silty soil. Wang et al. (2014) conducted indoor model test of a suction caisson penetrating silt under suction, and proposed a modified calculation method including the soil cohesion. Wang et al. (2018) investigated the effect of the installation suction and friction factor on the soil heave and resistance during the penetration in NC clay based on large deformation finite element analysis. Klinkvort et al. (2019) proposed a flow model to predict plug lift in layered soils. In practice, the applied suction should be controlled between the required and critical values. Two methods to predict the penetration resistance are introduced in this paper, which base on soil parameters and in-situ

CPT data respectively. Verification and calibration work of these methods are carried out against the monitoring data from the field installation of the mono-bucket installation in silty soil.

2 Prototype Field Test

The mono-bucket foundation consisted of a steel structure and a concrete transition piece, as shown in Fig. 2(a). In order to achieve a sufficient bearing capacity in weak sediment, the geometry dimension of the skirt had: Diameter $D = 30$ m, Length $L = 10$ m, Wall thickness $t = 0.025$ m. The circular bucket was divided into 7 sealed rooms by several inner walls (vertical stiffeners with a thickness $t = 0.02$ m), as shown in Fig. 2(b). The horizontal distance between the vertical stiffener and the skirt was around 8 m.

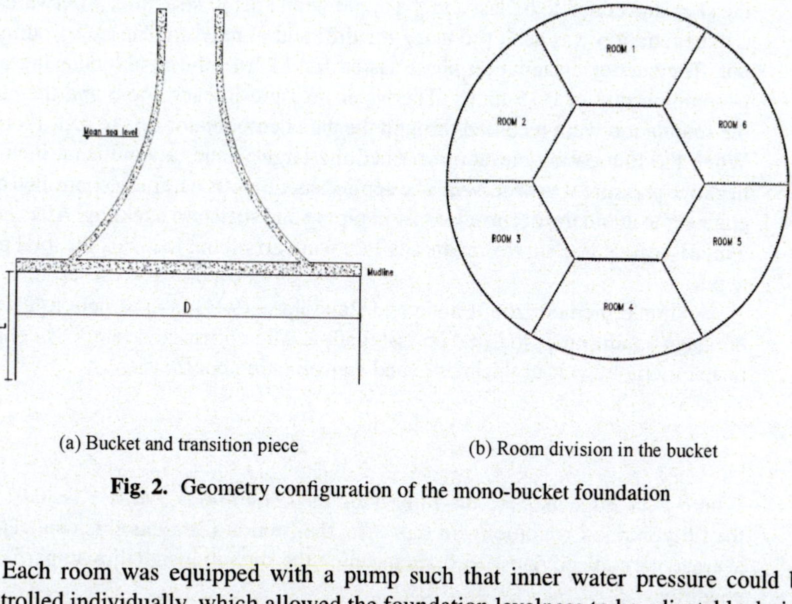

(a) Bucket and transition piece (b) Room division in the bucket

Fig. 2. Geometry configuration of the mono-bucket foundation

Each room was equipped with a pump such that inner water pressure could be controlled individually, which allowed the foundation levelness to be adjustable during the whole penetration process. Before the field installation, monitoring sensors were installed on the bucket including a water pressure sensor in each room and a gyroscope on the top of the foundation. A real-time kinematic GPS was set on the tower to precisely monitor the elevation of the mono-bucket during the penetration process. The mono-bucket would be installed into four layers of silty soil through self-weight penetration and suction penetration. The parameters of the four soil layers are listed in Table 1.

Table 1. Parameters of the four silty layers

Number	type	Thickness (m)	Submerged unit weight (kN/m^3)	Cohesion, c (kPa)	Friction angle (°)
Layer 1	Silty clay	2.5	6.5	5	15
Layer 2	Silt	2.5	7	13	22
Layer 3	Silt	4.0	8	9	28
Layer 4	Silty clay	3.0	7.8	7	19

By pumping air into the bucket, the mono-bucket can float in sea water and be shipped offshore by wet tow with a special tug. In the field, the foundation was lowered to the seabed by releasing the inside air, with the aid of a crane. At the beginning of installation, the skirt was penetrated into the seabed by 4 m under the dead weight of the skirt, the concrete ballast and the upper wind energy structure. Afterwards the other 6 m penetration was achieved using the differential pressure caused by pumping water out. The suction installation phase lasted for 11 h in the field, achieving an average penetration rate of 0.15 mm/s. The water pressure in each room and the elevation of the foundation were recorded through the data demodulator on the construction vessel. When the foundation elevation reached the target value, a significant increase in the negative pressure was observed. The applied suction was carefully controlled by the field engineer to avoid the occurrence of soil piping and structure buckling. After installation, a multi-beam sonar survey on the seafloor was carried out to check the total penetration depth.

A simple method from Finnie and Randolph (1994) is used here to determine the drainage condition of the suction installation. The normalized rate V is defined by the penetration rate v, wall thickness t and consolidation coefficient c_v:

$$V = \frac{vt}{c_v} \tag{1}$$

Where $V > 30$ stands for the fully undrained conditions, while $V < 0.01$ stands for the fully drained conditions. In this case, the suction installation is conducted with an average normalized rate $V = 0.13$, meaning the suction installation in partially drained condition.

During the dead-weight penetration, the downward driving force is equal to the self-weight of the whole structure minus the buoyancy in the water. During the suction penetration, the driving force can be calculated as the sum of the submerged dead-weight and the differential pressure between the inside the outside:

$$F_D = W' + \sum_{i=1}^{n} u_i A_i \tag{2}$$

where W' = the submerged weight of the structure, u_i = the underpressure in Room i, and A_i = the sectional area of Room i.

3 Penetration Resistance Prediction

Andersen et al. (2008) revealed that the penetration resistance of suction bucket foundation, P_f, consisted of the bearing capacity of the skirt tip, Q_{tip} and the friction force along the skirt wall, Q_{side}:

$$P_f = Q_{trip} + Q_{side} \tag{3}$$

For the hybrid mono-bucket foundation, the tip capacity Q_{tip} and the skirt friction Q_{side} should also include those both from the bucket skirt and inner stiffeners. The penetration resistance P_f can be calculated either from soil parameters from laboratory test or from in-situ cone penetration test (CPT) data. The first method is based on the bearing capacity theory with triaxial friction angle, in which undisturbed soil samples and precise laboratory tests are required. In the case of low permeability sediment (clay), the soil exhibits undrained behavior, where undrained shear strength should be used in the calculation. In the case of high permeability sediment (sand), drained shear strength and friction angle should be adopted to conduct an effective stress analysis (Klinkvort et al. 2019). The second method is empirically based on in-situ investigation (CPT) data, in which the coefficients are calibrated against the field experience and soil property.

4 Calculation Method Based on Soil Parameters

In this method, the tip capacity Q_{tip} and the wall friction Q_{side} are calculated individually from the friction angle in triaxial test:

$$P_f = Q_{tip} + Q_{side} = q_{tip}A_{tip} + A_{side}\int f_s(z)dz \tag{4}$$

where q_{tip} = the bearing capacity at the tip, A_{tip} = the tip area of the skirt wall and inner wall, A_{side} = the total area of the both sides on the skirt and inner stiffeners at current penetration depth, $f_s(z)$ = side friction along the wall at the depth of z. The resistance q_{tip} and f_s can be further described as follows:

$$q_{tip} = 0.5\gamma' t N_\gamma + qN_q \tag{5}$$

$$f_s(z) = K\sigma_v' \tan\delta = K\tan\delta\int\gamma' dz \tag{6}$$

$$N_q = e^{\pi\tan\phi}\tan^2\left(45 + \frac{\varphi}{2}\right) \tag{7}$$

$$N_\gamma = 1.5\left(N_q - 1\right)\tan\varphi \tag{8}$$

where t = the thickness of the wall, q = the effective stress at the skirt tip level, φ = the peak drained friction angle from the triaxial test, δ = the interface friction angle between soil and wall, K = the lateral earth pressure coefficient.

The interface friction angle can be expressed as $\delta = r\varphi$, where r was selected to be 0.9 with reference to Andersen et al. (2008). The lateral earth pressure coefficient K was recommended to be 0.8 in API (1993), as well as the analysis from the prototype cases, while a higher value of $K = 1.1$ was observed from model test results in Andersen et al. (2008).

The aforementioned equations are orginally introduced to predict the penetration resistance in cohesionless soil (e.g. sand). For silty sediment, modifications should be proposed to account for both the cohesion and the friction angle of the soil. Wang et al. (2014) introduced the cohesion c into the skirt friction, and the average friction $f_{s,av}$ is thus given as:

$$f_{s,\mathrm{av}} = 0.5\left(c + K\gamma' z \tan\delta\right) \tag{9}$$

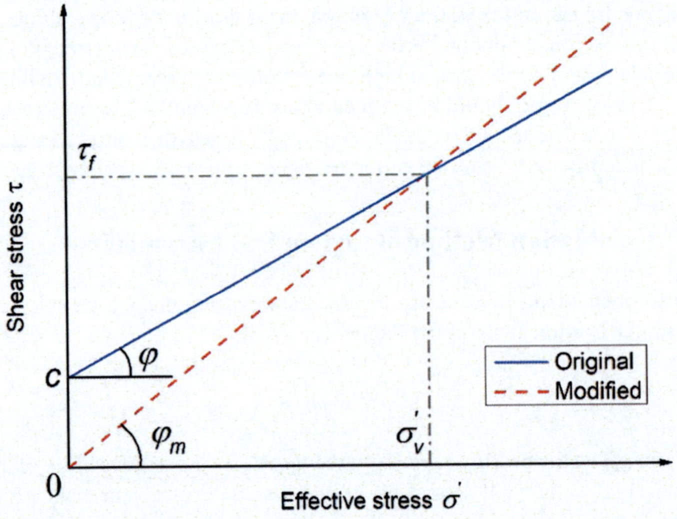

Fig. 3. Modification in friction angle φ

The other methodology proposed in this paper is to modify the friction angle φ following Mohr–Coulomb failure criterion. Under an effective stress of σ_v', a higher value of φ_m is adopted to represent the shear strength of the silt, instead of the cohesion c, as shown in Fig. 3. The shear strength can be calculated from the strength parameters c, φ and effective stress σ_v' as follows:

$$\tau_{\mathrm{f}}' = c + \sigma_v' \tan\phi = \sigma_v' \tan\phi_m \tag{10}$$

As the initial effective stress increases with depth, the modified friction angle φ_m also varies with the embedment depth even in the same soil layer.

5 Calculation Method Based on In-Situ CPT Test

Based on the in-situ CPT data, the penetration resistance can be calculated either only with cone tip resistance q_c or together with skirt friction f_s in Eq. (3) relating to the friction angle and the effective stress.

$$P_f = Q_{tip} + Q_{side} = k_{tip} q_c A_{tip} + A_{side} \int k_{side} q_c(z) dz \tag{11}$$

$$P_f = Q_{tip} + Q_{side} = k_{tip} q_c A_{tip} + A_{side} \int f_s(z) dz \tag{12}$$

where k_{tip} = empirical coefficient for tip resistance; k_{side} = empirical constant coefficient for wall friction.

The two empirical coefficients have significant effects on the calculation of the resistance P_f, and need to calibrated according to the local engineering experience. The recommended values of k_{tip} and k_{side} from DNV (1992) and Andersen et al. (2008) are listed in Table 2. It is noted that the friction coefficient k_{side} is much larger in clay than in sand, leading to the difficult in the determination of the coefficient k_{side} in silt. Except for the difference in soil type, the paint and polish on the steel-soil interface can also affect the side friction.

Table 2. Recommended values of k_{tip} and k_{side}

Reference	Soil type	k_{tip} value	k_{side} value
DNV (1992)	Clay	0.4	0.03
	Sand	0.3	0.001
Andersen (2005)	Mainly dense sand	0.01–0.55	0.0015
		0.03–0.60	0.001

6 Back-Analysis

Assuming the soil resistance is balanced by the driving force: $F_D = P_f$ in the suction penetration phase, the penetration resistance can be calculated using the underpressure data from monitoring. During self-weight penetration, positive pressure was also monitored within the bucket, which was be used to calculate the resistance.

Figure 4 compares the prediction of penetration resistance using Eq. (3–6) based on soil parameters with the prototype monitoring data. In the modified calculation method, the modified friction angle φ_m from Eq. (8) is used instead of the original friction angle φ. It is demonstrated that the curve with modified method is closer to the monitoring result, while the prediction with the original friction angle underestimates the soil resistance. At the end of installation, the resistance increased rapidly because the bucket lid was stopped by the inner soil plug.

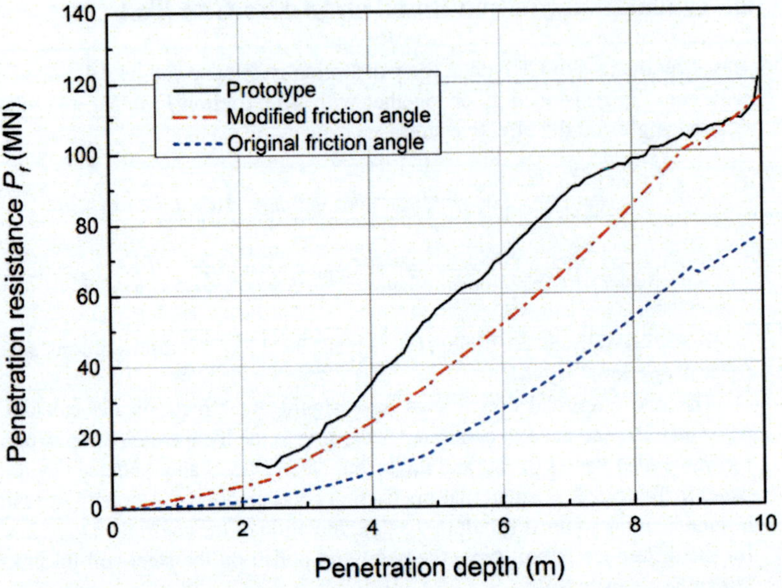

Fig. 4. Prediction of penetration resistance using soil parameters ($K = 1.05$)

The lateral earth pressure coefficient K is set to be 1.05 to give a better fitting against the field data, which is larger than the recommend value of 0.8 by Andersen et al. (2008) and API (1993). This can be attributed to the different soil type and the existence of the stiffeners. Assuming the soil is normally consolidated, the earth pressure coefficient at rest K_0 is normally larger in clay or silt than the value in sand, which leads to a larger initial lateral stress in prediction. The inner vertical stiffeners in mono-bucket occupied a part of volume of the suction bucket, and caused the compression of the surrounding soil. They may be the reasons to adopt a higher K for the hybrid mono-bucket in silt than the calibration results in dense sand from Andersen et al. (2008).

In the second method, the original in-situ CPT data is used to calculate the resistance by Eq. (9). Calibration work of the two empirical coefficients k_{tip} and k_{side} is carried out against a series of prototype monitoring data, as shown in Fig. 5. And the corresponding coefficients k_{tip} and k_{side} from calibration are listed in Table 3. The calculation result treating silt as clay is significantly higher than the resistance from prototype, while the result treating silt as sand presents the opposite trend.

The penetration resistance prediction using the second method using CPT data shows a better agreement with the prototype monitoring data, meaning it is more reliable to use in-situ CPT data rather than soil parameters. The possible reason may be that the CPT data represents the undisturbed soil condition, but the soil parameters from laboratory may be influenced by the sampling disturbance and the artificial judgement of the technician.

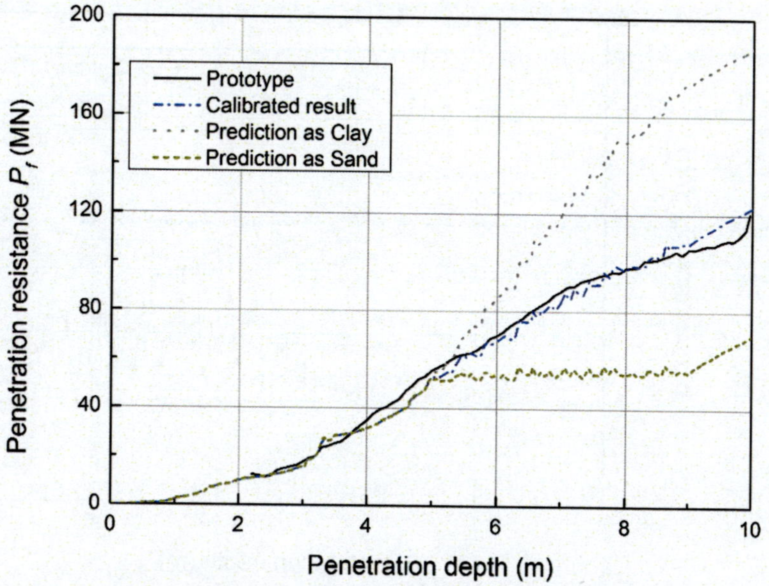

Fig. 5. Prediction of penetration resistance using in-situ CPT data

Table 3. Empirical coefficient k_{tip} and k_{side} from calibration.

Number	Type	k_{tip}	k_{side}
Layer 1	Silty clay	0.4	0.04
Layer 2	Silt	0.35	0.015
Layer 3	Silt	0.35	0.015
Layer 4	Silty clay	0.4	0.04

For pure silt, it is recommended to set the coefficients k_{tip} and k_{side} to be the mean values of the coefficients for clay and sand from DNV (1992) in Table 2, i.e. $k_{tip} = 0.35$, and $k_{side} = 0.015$. However, for silty clay, it is recommended to determine the two coefficients regarding the plasticity index and other soil properties.

A comparison between the calculated tip resistance Q_{tip} is shown in Fig. 6. Two curves are obtained by the first method using the original friction angle φ and the modified friction angle φ_m. The other curve is calculated by the second method using CPT data. A good fit is found between the curves using the original friction angle φ and CPT data against the field monitoring results, while using the modified firction angle φ_m may overestimate the tip bearing capacity in shallow soil.

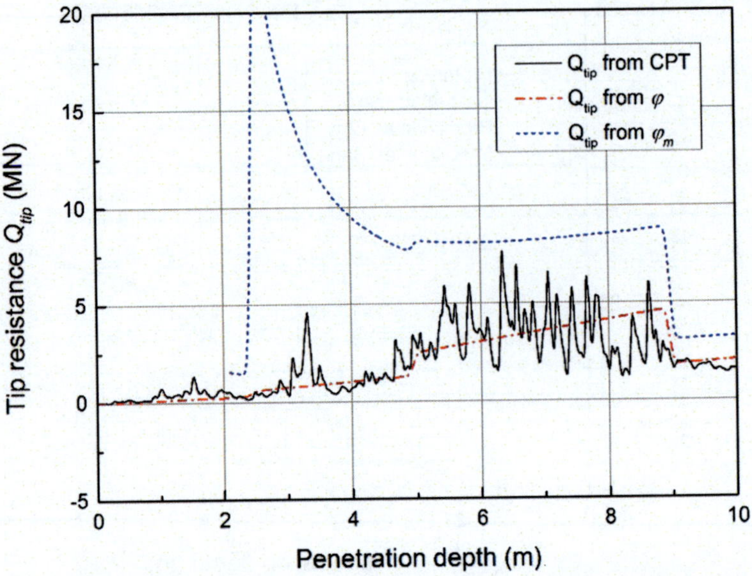

Fig. 6. Comparison of the tip resistance Q_{tip} between two methods

7 Discussion

The observation reveals that the soil resistance grows rapidly at the penetration depth of 9.8 m, 0.2 m less than the skirt length, which is probably caused by the soil heave from suction installation. Due to the small L/D ratio of the mono-bucket in the field, it is assumed the distance between the inner stiffeners and the skirt (8 m) is large enough to ignore the increase in soil stress from penetration.

In this field work, the quality of the soil sample in shallow depth may not be in its best quality due to the difficulty in sampling and transportation from offshore boreholes. The above-mentioned method using soil parameters greatly relies on the quality of in-situ sampling and laboratory test, and the geotechnical interpretation of the test data.

Efforts will be made both in the field monitoring and numerical analysis to evaluate the effect of the flow condition in silty soil. The calibration work of the prediction methodologies is still undergoing as more mono-bucket foundations and suction bucket supported jackets are being designed and constructed in those areas with silty sediment.

8 Conclusion

The penetration resistance prediction is critical to the design and application of the hybrid mono-bucket. Two calculation methods, based on soil parameters and in-situ CPT data respectively, are calibrated against the prototype installation monitoring data of 11 mono-buckets in silty soil. Some conclusions are drawn as follows:

(1) The first method based on soil parameters can be used to calculate the total penetration resistance in silty soil by Eq. (4–8), with a modification in the interface friction angle to account for the cohesion of silt acting on the skirt.
(2) A lateral earth pressure coefficient of $K = 1.05$ is calibrated against the field monitoring data in silty soil, which is greater than the recommended value of 0.8 from the experience in the dense sand of North Sea.
(3) The second method based on in-situ CPT resistance by Eq. (11) is verified with prototype monitoring data. It can be more precise and reliable than the first method, with appropriate empirical coefficients k_{tip} and k_{side}.
(4) The empirical coefficients k_{tip} and k_{side} should be decided according to the local experience and soil types before application. For pure silt sediment, it is recommended to adopt the mean value of the coefficients in sand and clay from API (1993), $k_{tip} = 0.35$, and $k_{side} = 0.015$. But the coefficients in soft silty clay can be varying from its property.

Acknowledgement. The authors are grateful to the support from Zhejiang Provincial Natural Science Foundation of China (LQ19E090002 and LQ20E090001) and Postdoctoral Science Foundation of China (2018M642383). Discussion with Knut H. Andersen and other colleagues from Norwegian Geotechnical Institute (NGI), field work from Jiangsu Daoda Engineering and the help from Tianjin University and Zhejiang University are also highly appreciated.

References

American Petroleum Institute: API. RP 2A-LRFD. Planning Designing and Construction of Fixed Offshore Plat-forms—Load and Resistance Factor Design. American Petroleum Institute (1993)

Andersen, K.H., Jostad, H.P., Dyvik, R.: Penetration resistance of offshore skirted foundations and anchors in dense sand. J. Geotech. Geoenviron. Eng. **134**(1), 106–116 (2008)

Andersen, K.H., et al.: Suction anchors for deepwater applications. In: Proceedings of the 1st International Symposium on Frontiers in Offshore Geotechnics, ISFOG, Perth, pp. 3–30, September 2005

Byrne, B.W., Houlsby, G.T.: Foundations for offshore wind turbines. Philos. Trans. R. Soc. Lond. Ser. A Math. Phys. Eng. Sci. **361**(1813), 2909–2930 (2003)

Det Norske Veritas: DNV CN 30.4. Foundations (1992)

Ding, H., Liu, Y., Zhang, P., Le, C.: Model tests on the bearing capacity of wide-shallow composite bucket foundations for offshore wind turbines in clay. Ocean Eng. **103**, 114–122 (2015)

Erbrich, C.T., Tjelta, T.I.: Installation of bucket foundations and suction caissons in sand-geotechnical performance. In: Offshore Technology Conference (1999)

Finnie, M.S., Randolph, M.F.: Punch-through and liquefaction induced failure of shallow foundations on calcareous sediments. In: Proceedings of International Conference on Behaviour of Offshore Structures, Boston, MA, pp. 217–230 (1994)

Guo, Z., Wang, L., Yuan, F., Li, L.: Model tests on installation techniques of suction caissons in a soft clay seabed. Appl. Ocean Res. **34**, 116–125 (2012)

Houlsby, G.T., Byrne, B.W.: Design procedures for installation of suction caissons in clay and other materials. Proc. Inst. Civil Eng. Geotech. Eng. **158**(2), 75–82 (2005a)

Houlsby, G.T., Byrne, B.W.: Design procedures for installation of suction caissons in sand. Proc. Inst. Civil Eng. Geotech. Eng. **158**(3), 135–144 (2005b)

Klinkvort, R.T., Sturm, H., Andersen, K.H.: Penetration model for installation of skirted foundations in layered soils. J. Geotech. Geoenviron. Eng. **145**(10), 04019085 (2019)
Nielsen, S.D., Ibsen, L.B., Nielsen, B.N.: Dynamic behaviour of mono bucket foundations subjected to combined transient loading. In: The 3rd Internationl symposium on Frontiers in offshore Geotechnics, pp. 313–318. CRC Press, May 2015
Randolph, M., Gourvenec, S.: Offshore Geotechnical Engineering. CRC Press, Boca Raton (2017)
Senders, M., Randolph, M.F.: CPT-based method for the installation of suction caissons in sand. J. Geotech. Geoenviron. Eng. **135**(1), 14–25 (2009)
Sturm, H.: Design aspects of suction caissons for offshore wind turbine foundations. In: Proceedings of the 19th International Conference on Soil Mechanics and Geotechnical Engineering: Unearth the Future, Connect Beyond (2017)
Tjelta, T.I.: The suction foundation technology. Front. Offshore Geotech. III, 85 (2015)
Tran, M.N., Randolph, M.F., Airey, D.W.: Installation of suction caissons in sand with silt layers. J. Geotech. Geoenviron. Eng. **133**(10), 1183–1191 (2007)
Wang, L., Yu, L., Guo, Z., Wang, Z.: Seepage Induced soil failure and its mitigation during suction caisson installation in silt. J. Offshore Mech. Arct. Eng. **136**(1), 011103 (2014)
Wang, Y., Zhu, X., Lv, Y., Yang, Q.: Large deformation finite element analysis of the installation of suction caisson in clay. Mar. Georesour. Geotechnol. **36**(8), 883–894 (2018)
Zhang, P., Han, Y., Ding, H., Zhang, S.: Field experiments on wet tows of an integrated transportation and installation vessel with two bucket foundations for offshore wind turbines. Ocean Eng. **108**, 769–777 (2015)

CPT and SPT as Complementary Tests for the Formulation of Geotechnical Design Profiles

Emmanouil Spyropoulos[1](✉), John Fahd Touma[2], Habib-ur-Rehman Ahmed[2], and Muhammad Abdul Waheed[2]

[1] Saudi Aramco, Dhahran, Saudi Arabia
emmanouil.spyropoulos@aramco.com
[2] Riyadh Geotechnique and Foundations (RGF), Riyadh, Saudi Arabia
{j.touma,habib.rehman,abdul.waheed}@erigeosciences.com

Abstract. Standard Penetration Test (SPT) is the most commonly used field test in any geotechnical investigation campaign. Although SPT provides quite useful information on the nature and type of the subsurface strata, owing to its inherent operating mechanism, SPT incorporates several limitations. During the dynamic impact of the SPT hammer during the testing/sampling, there is a tendency of the breakage of the cementation and disturbance of the structure in the partially-cemented soils. Breakage of cementation and similar disturbance to the soil structure may mask some of the essential information required for the choice of geotechnical design parameters for the in-situ strata. Such partially-cemented soils are prevalent in the semi-arid regions of the world such as eastern Saudi Arabia. Cone Penetration Test (CPT) is another penetration-based field test employed, usually, deployed as a secondary test for the geotechnical explorations. Due to the pseudo-static penetration mechanism of the cone during CPT testing, this test is capable of quite closely recording the information pertinent to the undisturbed structure of the soil. Therefore, CPT can be classed as an essential complementary test to SPT for the development of the representative in-situ design subsurface profile. This paper discusses several case studies from eastern Saudi Arabia, in which the precise interpretation of the CPT provided the essential complementary information that augmented the results from the SPT in developing the accurate and field representative subsurface design profiles.

1 Introduction

In any routine geotechnical investigation campaign, number of field testing techniques are used. All these techniques have their corresponding pros and cons. Standard penetration test (SPT) is being widely utilized to delineate nature and type of subsurface strata. Many geotechnical design parameters of the soils are associated with the standard penetration test (SPT) and huge database of SPT correlations is available to assess these design parameters. Cone Penetration Test (CPT) is another penetration-based field test employed, usually, as a secondary test for the geotechnical explorations. CPT has several advantages over SPT including continuous profiling, relatively quick performance and

J. J. Yang et al. (Eds.): GeoChina 2021, SUCI, pp. 83–92, 2021.
https://doi.org/10.1007/978-3-030-79672-3_6

less/no disturbance affects associated with boreholes. On contrary, its limitations include non retrieval of samples and limited penetration in very dense and hard materials. The combined results of both of these tests however, could provide quite essential information about the subsurface strata, which each one of these in isolation may not be able to reveal.

SPT and CPT have been correlated in literature as $n = q_c/N_{60}$ where q_c is CPT cone bearing resistance and N_{60} is corrected SPT N-value. SPT and CPT correlations are generally used for subsurface strata classification and translation of SPT blow counts (N-values) into CPT cone resistances (q_c-values) and vice versa. The primary objective of conversion of one parameter into the other is to use the available data base of correlations in literature depending upon the local experience, field observations, design methodologies and preferences.

De Alencar Velloso (1959) were among the first who determined ratios of SPT N values and cone tip resistance for different soil types and introduced 0.35 MPa for clay and silty clay, 0.2 MPa for sandy clay and silty clay, 0.35 MPa for sandy silt, 0.6 MPa for fine grained sand, and 1.0 MPa for sand. Meigh and Nixon (1961) considered the effect of grain size in correlation proposed by De Alencar Velloso (1959) and recommended n-values as 0.2 MPa for coarse grained sand and 0.3–0.4 MPa for gravelly sand. Schmertmann (1970) recommended 0.2 MPa for silt, sandy silt, and slightly cohesive silt-sand mixture, 0.3–0.4 MPa for clean, fine to medium sand and slightly silty sand, 0.5–0.6 MPa for sandy gravel and gravel. Meyerhof (1965) suggested the relationship as $n = (q_c/N) = 0.4$ MPa. Robertson et al. (1983) used the available data base in literature in combination with his filed testing in the province of British Columbia in Canada and developed D_{50} based N_{60} and CPT correlation. Robertson et al. (1986) later normalized cone tip resistance (q_c/Pa) and presented a CPT-SPT correlation for different soil types. Kulhawy and Mayne (1990) extended the Robertson et al. (1983) correlation based on their additional data and developed a mathematical expression for their updated SPT-CPT correlation based on fines content. Lunne et al. (1997) modified the CPT-SPT correlation by Robertson et al. (1986) by using Soil Behavior Index (I_c).

Ismael and Jeragh (1986) made a correlation on calcareous desert sands in Kuwait and compared it with the value of Schmertmann (1970) for clean, fine to medium sands and slightly silty sands. Their average value ranges from 4.2 to 5.6 which was higher than what Schmertmann suggested for clean, fine to medium sands and slightly silty sands. N. Akca (2003) carried out study on United Arab Emirates soils (Sand, Silty Sand, Sandy Silt) and results of q_c/N_{60} ratio are generally higher than literature values. The deviation of the q_c/N_{60} ratio in both of these studies from the ones available in the literature was attributed to cementation, densification, gravels/shells and calcareous nature of soils.

Table 1 represents the summary of correlative works which have been done in the past to develop correlations between SPT N-values and CPT cone bearing resistances, q_c.

The soils found in eastern Saudi Arabia, especially in coastal areas are generally cemented and calcareous in nature with shallow water tables. Therefore, it would be useful to study the correlations between SPT N-values and CPT cone bearing resistances (q_c) for local soils to get an insight into subsurface strata behaviour/response in both tests.

Table 1. q_c/N_{60} ratios available in literature (q_c in MPa).

Soil Type	De Alencar Velloso (1959)	Schmertmann (1970)	Engineers Franki Piles (1960)	Meigh and Nixon (1961)	Meyerhof (1965)	Robertson et al. (1986)	Danziger and De Valleso (1995)	Acka (2003)	Ismael and Jeragh (1986)
Sand	0.6–1.0	0.3–0.6	1.0	0.20	–	0.3–0.6	0.3–0.6	0.77	4.2 to 5.6*
Silty Sand	0.2	0.3–0.4	0.5	–	0.4	0.3–0.4	0.3–0.4	0.7	–
Sandy Silt	0.35	0.2	0.4	–	–	–	0.2	0.58	–

* q_c in kgf/cm^2.

2 Sites and Investigation Description

The data set used in this study consists of Cone Penetration Test (CPT) soundings and standard penetration test (SPT) profiles from three (03) different sites in eastern Saudi Arabia. All data used in this study are from the coastal areas of eastern Saudi Arabia. Location map of the sites is shown in Fig. 1. Data set from each site consists of five (05) boreholes and five (05) CPTs. CPTs at each site were carried out at about 3.0 to 5.0 m distance from the SPT boreholes.

A total of fifteen (15) boreholes with SPTs were drilled at the site locations. All the boreholes were drilled using straight rotary drill rigs. Standard Penetration Tests (SPTs) were performed using automatic hammers at 0.75 m depth intervals. These tests were performed generally in accordance with ASTM D-1586 using a split spoon sampler of 35 mm inner dia. and 50 mm outer diameter.

CPT soundings were also performed at fifteen (15) locations in the vicinity of boreholes. CPTs were performed using Geomil electric cone penetrometer. In order to perform the CPT, ten (10) tons capacity wheel mounted hydraulic rig was used. The tests were performed in accordance with procedures described in ASTM designations D5778. A calibrated electric cone is pushed into the ground by hydraulic pressure. The electric cone connected to a microprocessor and a laptop computer passed signals thru a cable regarding the tip resistance, side friction as well as pore water pressure measured by the cone. All data is automatically recorded in the laptop and later processed and printed using patented software.

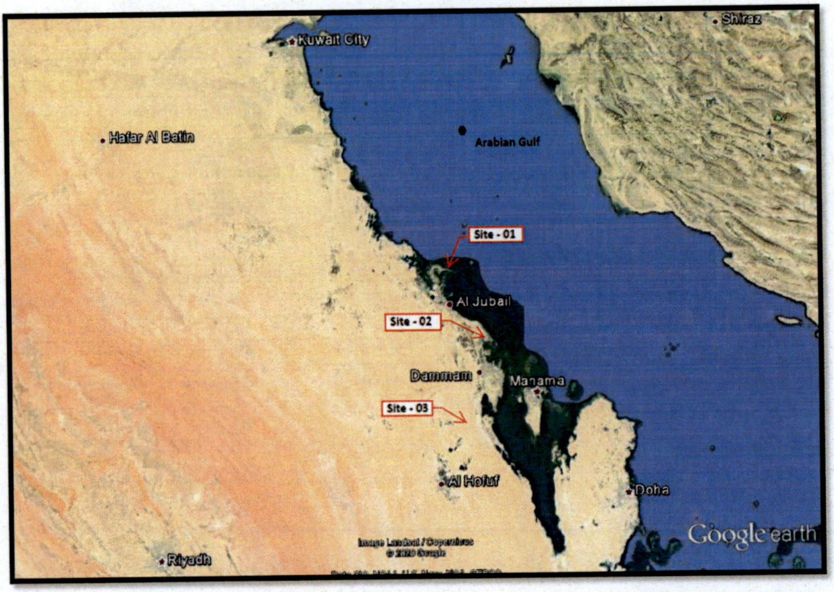

Fig. 1. Vicinity Map showing location of three (03) investigated sites

2.1 Subsurface Strata Conditions

A particular feature, typical for the western and southern Gulf coast, is the flat and wide spreading coastal Sabkha. Penetrating up to 10 km inland they cover areas of more than 100 km^2. "Sabkha" is the Arabic term for flat salt-crusted desert. Sabkha is a saline (generally coastal) deposit consisting of saturated, loose silty sand and possibly clay. Sabkhas are salt bearing arid climate sediments (Saudi Building Code 303). In general, Sabkha sediments are characterized by high void ratios and low dry densities. Accordingly, upon wetting Sabkha soil is renowned for being highly compressible material with low bearing resistance, and hence considered among the poorest of foundation materials.

Soil layers consisting of Poorly Graded SAND (SP)/Poorly Graded SAND with Silt (SP-SM)/Silty Sand (SM)/Silt (ML) have been predominantly found at borehole locations in the study areas. Very loose/loose layers encountered in boreholes and CPTs are classified as sabkha layers. Being very near to shoreline, the Sabkha layers found at the study sites are directly connected to sea water and exhibit characteristics of "coastal Sabkha".

The groundwater table was encountered in the boreholes at shallower depths varying from 1.50 to 7.0 m below existing grade level. As the groundwater table is connected to sea, it has very high dissolved salt content and may experience seasonal/tidal fluctuations.

3 SPT and CPT Data Analysis

For all the boreholes of the three investigated sites, SPT-N values were normalized to N_{60} after applying the energy and other essential corrections. The tip resistance (q_c) values were also logged from the CPT profiles corresponding to the depths at which SPTs were performed. It is also necessary to normalize the cone resistance for pore pressure.

From the fifteen (15) SPT boreholes and CPTs, the number of data pairs (N_{60} and q_c) available for comparison are about one hundred and fifty (150). However, just to explain the process, the data points for sands where sieve analysis information was also available have been selected and tabulated in Table 2.

Table 2. q_c/N_{60} field data points.

Depth	Fines content (%)	D_{50} (mm)	N_{field}	N_{60}	q_c (MPa)	$n = q_c/N_{60}$ (MPa)
7.50	6.60	0.28	23	32	17.56	0.55
15.0	23.40	0.10	32	47	12.83	0.28
0.75	1.70	0.28	7	7	9.01	1.27
9.0	10.20	0.22	17	22	4.34	0.20
3.0	7.40	0.35	7	7	26.94	3.80
3.75	8.20	0.28	8	8	7.70	0.95
1.50	10.50	0.25	9	9	4.30	0.47
4.50	5.60	0.26	22	25	15	0.59

Finally for all these boreholes, q_c/N_{60} ratios were plotted against depth below the ground surface and are shown in Figs. 2, 3 and 4. The soil samples recovered from SPT split-spoon samplers were classified through laboratory-based particle size distribution and the plasticity tests. Based on the type of soils classified from the SPT samples, a range of q_c/N_{60} ratios, based on the correlations available in the literature, were assigned to these soils and plotted on the same figures (Figs. 2, 3 and 4). Most commonly adopted relationships developed by Schmertmann (1970) and Robertson et al. (1986) have been used to assign the range of q_c/N_{60} ratios to the corresponding soils in these plots.

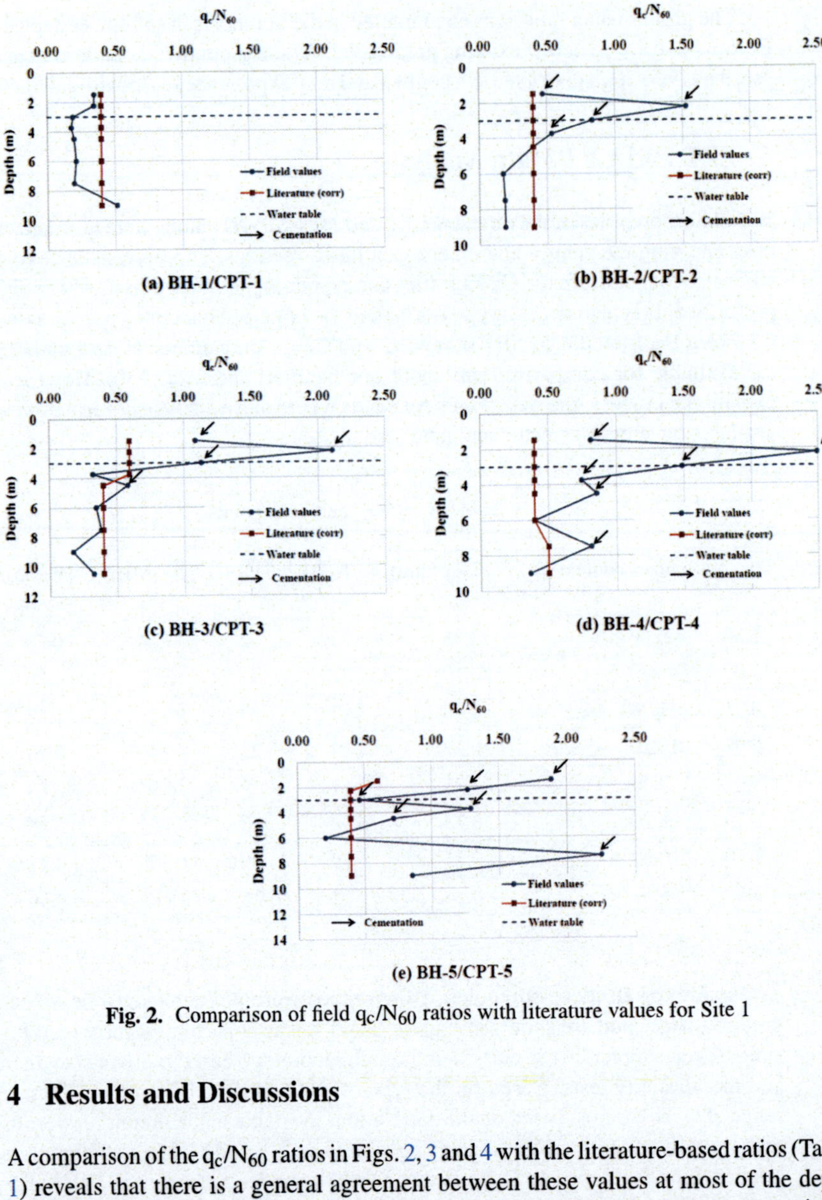

Fig. 2. Comparison of field q_c/N_{60} ratios with literature values for Site 1

4 Results and Discussions

A comparison of the q_c/N_{60} ratios in Figs. 2, 3 and 4 with the literature-based ratios (Table 1) reveals that there is a general agreement between these values at most of the depth horizons with some exceptions especially above the water table. At these exceptional locations, actual (field) ratios have been found to be quite higher than the ones generally available in the literature. Moreover, almost all the places having a general disagreement between the field and literature-based ratios lie above the water table. As discussed in the site description section, the investigated sites are typified by the shallow water tables and very high dissolved salt content. The high salt content of these waters owes

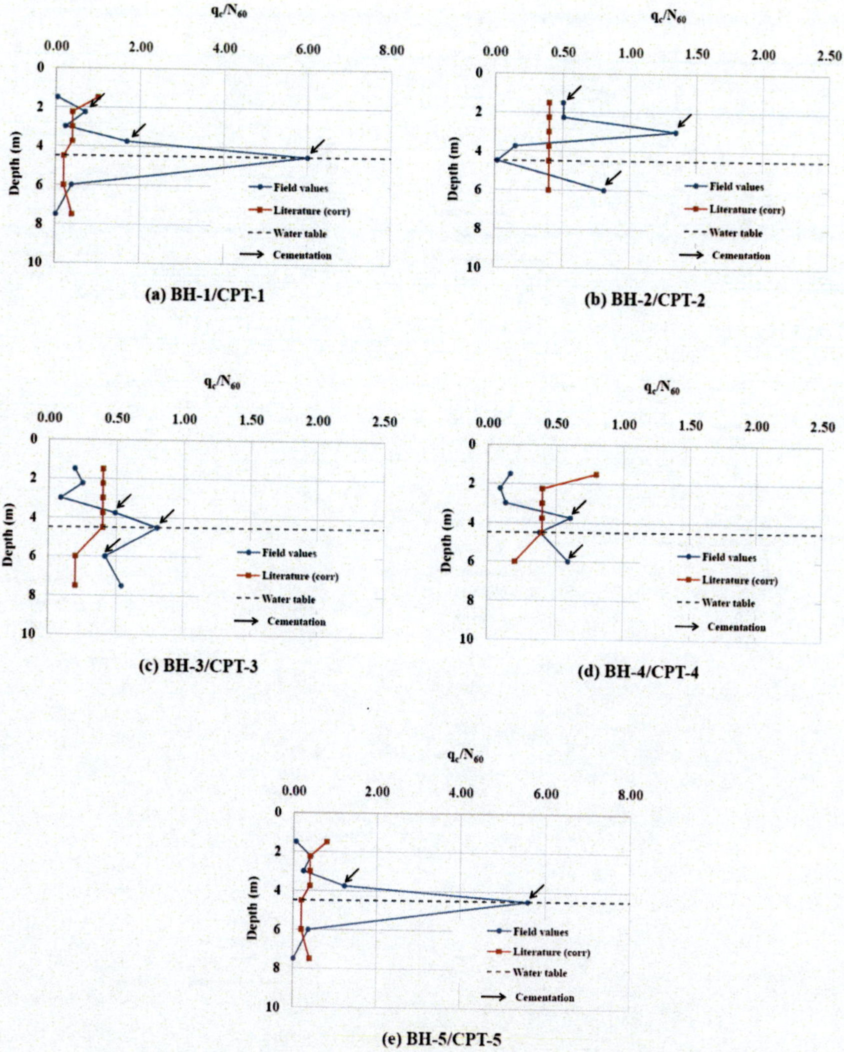

Fig. 3. Comparison of field q_c/N_{60} ratios with literature values for Site 2

either to the presence of percolated seawater or to the Sabkha brines. Due to several factors including fluctuation of the water table with tidal variation, ambient temperature, relative humidity, and precipitation, the soils above the water table experience continuous cycles of capillary saturation and evaporation. These cycles result in the precipitation and the consequent cementation of different types and amounts of salts in the soil layers above the water table. This cementation process has been confirmed by the undisturbed samples acquired through coring during several investigation campaigns in the area. The

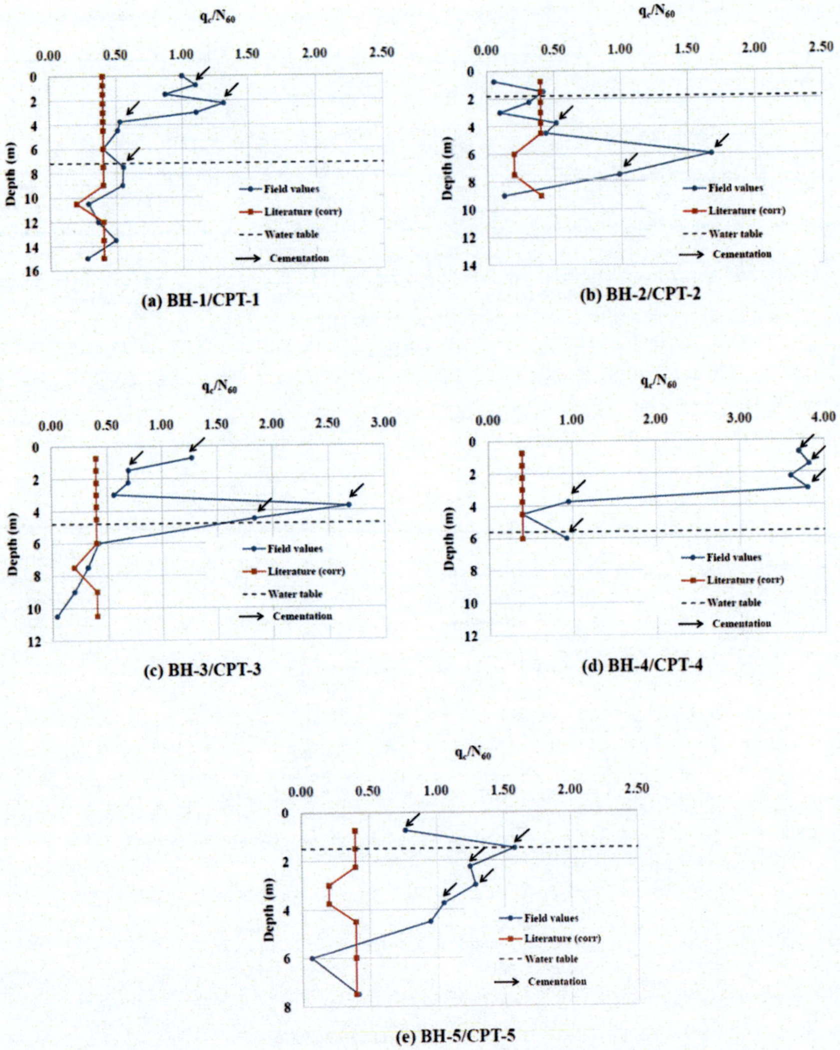

Fig. 4. Comparison of field q_c/N_{60} ratios with literature values for Site 3

cementation among the soil particles, most probably, results from the precipitation of salts such as calcium carbonate, sodium chloride, potassium chloride, etc.

During any regular field investigation campaign using SPT on a site with cemented subsurface layers, the dynamic impact of SPT hammer results in the breakdown of the cementation among the soil particles. After the breakdown of the cementation, the SPT test proceeds virtually in an 'uncemented soil,' and the resulting N-values falsely represent the real/in-situ shear strength and stiffness behavior of these soils. CPT, on the other hand, due to its push-type and pseudo-static rate of penetration mechanism, provides a true representation of the nature and type of the cemented strata. Therefore, the deviation of the qc/N_{60} ratio determined at specific depth horizons in our projects from the ones available in the literature, most probably owes to the presence of the soil cementation at these depths. The degree of deviation of this ratio from the correlations in the previous research could also be a qualitative indicative of the extent of the cementation. In addition to the assessment of the cemented strata above the water table, this ratio could also be helpful in revealing the presence of such layers below the water table. In the seashore areas, presence of coralline or other such cemented deposits even below the water table could also be explored/revealed by the adoption of both SPT and CPT testing regimes. In our case, higher q_c/N_{60} ratios below the water table at some boreholes are also quite probably indicative of the cemented strata.

The findings of the current study have also been substantiated by similar studies on the calcareous/cemented soils in the Middle East. The ratio $(q_c/Pa)/N_{60}$, as determined from field data, for the UAE calcareous fine to medium grained sand was found to be significantly, 40 to 45%, higher than the values predicted using the existing correlations for siliceous sand (Elkateb and Ali 2010). They further highlighted that the existing correlations in literature are developed for siliceous sands and therefore have poor applicability in calcareous soils due to their high compressibility, crushability and presence of shells. Their study also indicated that the ratio (qc/Pa)/N60 for calcareous sand is higher than that of siliceous sand due to presence of shells in calcareous sands which might have a bigger influence on the CPT tip resistance value compared to the SPT blow counts. Ahmed et al. (2013) also concluded that none of the existing SPT-CPT correlations are applicable to calcareous sands and there is a need for reliable SPT-CPT correlations, encompassing both calcareous and siliceous soils. Moreover, the SPT-CPT correlation should address the anticipated difference in soil crushability/compressibility between the CPT and the SPT in calcareous soils.

5 Conclusions

Owing to their push-type and pseudo-static soil penetration methodology, CPT test have been found quite suitable to provide at least a qualitative assessment of the degree of the cementation present in the strata both above and below the water table.

The combined results from both SPT and CPT tests were able to provide quite essential information about the subsurface strata especially in cemented strata, which each one of these in isolation may not be able to reveal.

6 Recommendations

For the foundation design purposes, if the breakdown of the cementation among the soil particles is anticipated under the envisaged structure loading, the choice of geotechnical design parameters should conservatively be based on SPT-N values. Use of CPT values for the bearing capacity analysis in such cases could result in unsafe designs. However, in cases where there is a high degree of cementation in the strata and the probability of its persistence during the envisaged structure loading is high, the selection of foundation design parameters should be based on the CPT results. Use of SPT values in the latter case would result in uneconomical foundation designs.

7 Limitation

Although, assessment of the cementation in soil layers were carried out from the undisturbed samples acquired at different depths, quality of undisturbed samples may be further improved to refine the q_c/N ratios.

References

Meigh, A.C., Nixon, I.K.: Comparison of in-situ tests of granular soils. In: Proceedings of 5th International Conf. on Soil Mechanics and Foundation Engineering, Paris, France (1961)

Agaiby, S., Abdel-Rahman, A., Ahmed, S., Baghdady, A.: q_c/N ratio for recently deposited crushable soils. In: International symposium on advances in foundation engineering (ISAFE 2013) 5–6 December 2013

Akca, N.: Correlation of SPT-CPT data from the United Arab Emirates. Eng. Geol. **67**, 219–231 (2003)

Chin, C.T., Duann, S.W., Kao, T.C.: SPT-CPT correlations for granular soils. In: Proceedings of 1st International Symposium on Penetration Testing, vol. 1 pp. 335–339

Danziger, B.R., Velloso, D.A.: Correlations between the CPT and the SPT for some Brazilian soils, Proc. CPT 1995. Linkoping **2**, 155–160 (1995)

De Alencar Velloso, D.: O ensaio de diepsondeering e a determinacao da capacidade de carga do solo. Rodovia **29**, 1959 (1959)

Douglas, B.J., Olsen, R.S.: Soil classification using electric cone penetrometer. In: American Society of Civil Engineers, ASCE, Proceedings of Conference on Cone Penetration Testing and Experience, St. Louis, October 26–30, pp. 209–227 (1981)

Fellenius, B.H., Eslami, A.: Soil profile interpreted from CPTu data. Geotechnical Engineering Conference, Asian Institute of Technology, Bangkok, Thailand (2000),

Meyerhof, G G.: Shallow foundations. J. Soil Mech. Found. Div. ASCE, **91**(SM2), 21–31 (1965)

Ismael, N.F., Jeragh, A.M.: Static cone tests and settlement of calcareous desert sands. Can. Geotech. J. **23**(3), 297–303 (1986)

Lunne, T., Robertson, P.K., Powell, J.J.M.: Cone Penetration Testing in Geotechnical Practice. Blackie Academic, p. 312. EF Spon/Routledge Publ., New York, (1997)

Robertson, P.K., Campanelle, R.G.: Interpretation of cone penetration tests. Part I: Sand. Canadian Geotechn. J. **20**(04) (1983)

Robertson, P.K., Campanelle, R.G.: Interpretation of cone penetration tests. Part II: Clay. Canadian Geotechn. J. **20**(04) (1983)

Robertson, P.K., Campanelle, R.G.: SPT-CPT Correlations. ASCE, J. Geotech. Eng. **109**(11), 1449–1459

Schmertmann, J.H.: Guidelines for Cone Penetration Test, Performance and Design. U.S. Department of Transportation, Report No. FHWA-TS-78–209, Washington, D.C., p.145 (1978)

Ahmed, S.M., Agaiby, S.W., Abdel-Rahman, A.H.: A unified CPT–SPT correlation for non-crushable and crushable cohesionless soils. Ain Shams Eng. J. (2013)

Elkateb, T.M., Ali, H.E.: CPT-SPT correlations for calcareous sand in the Persian Gulf area. In: 2nd International Symposium on Cone Penetration Testing, Huntington Beach, CA, USA (2010)

Kulhawy, F.H., Mayne, P. W.: Manual on estimating soil properties for foundation design. Report EL-6800, Electric Power Research Institute, Palo Alto, California, USA (1990)

Research on Rapid Detection Technology Based on Infrared Image and Calibration Method of Tunnel Cracks

Yilin Wu[1], Yanlong Zhang[1,2], Qingyan Tian[1,2], and Haishan Lin[1,2](✉)

[1] Guangdong Hualu Transport Technology Co., Ltd., Guangzhou 510550, Guangdong, PR China
[2] Guangdong Jiaoke Testing Co., Ltd., Guangzhou 510550, Guangdong, PR China

Abstract. Because of the advantages of fast detection speed and no need for traffic control, the rapid detection vehicle based on camera technology is more suitable for the detection of apparent diseases in high-grade highway tunnels. However, this type of rapid detection vehicle basically uses visible light illumination, which has defects such as interfering with the normal driving of other vehicles in the tunnel. Therefore, a rapid detection vehicle for tunnel cracks using infrared supplementary light and its tunnel crack measurement technology based on digital images are introduced. Furthermore, in order to ensure the accuracy of crack detection of fast detection vehicle, it is necessary to calibrate the accuracy. In this paper, a new calibration method is proposed for the shortcomings of the existing calibration methods crack of detection accuracy. As an example, the rapid tunnel inspection vehicle based on the infrared band imaging technology is calibrated and compared with the other tunnel rapid inspection vehicles using visible light illumination. It is found that the calibration method is suitable for the calibration of the crack detection accuracy of tunnel rapid inspection vehicles with infrared or visible light illuminated. It also has the advantages of conforming to the actual environment of the operating tunnel, not affecting the traffic operation, small image processing workload, high accuracy, repeatable calibration, etc., and is further verified and reliable by the operating road tunnel, which meets the requirements of the relevant specifications for the accuracy of crack detection.

Keywords: Highway tunnel · Infrared fill light · Cracks · Rapid detection · Detection accuracy · Calibration method

1 Introduction

The development of tunnel rapid detection technology and equipment is more mature, and related products have been widely used in the rapid detection of practical projects. Detection methods are developing, such as impact echo nondestructive testing technology [1, 2], automatic detection technology based on camera technology [3–5], etc. According to different types of sensors, tunnel rapid detection technology can be divided into laser scanning technology and camera photography technology [6], in which laser

J. J. Yang et al. (Eds.): GeoChina 2021, SUCI, pp. 93–110, 2021.
https://doi.org/10.1007/978-3-030-79672-3_7

scanning technology uses vehicle laser scanner to scan and record the point cloud data on the inner surface of the tunnel in all directions, and then forms gray-scale image through post-processing software, based on which to identify the apparent diseases in the tunnel. The acquisition speed of this technology is generally less than 20 km/h [7–9]. Based on the camera technology, the vehicle mounted high-resolution camera and auxiliary lighting are used to collect and store the image of the inner surface of the tunnel in a fast and continuous manner. The gray-scale image is analyzed through the post-processing algorithm program, and then the apparent diseases such as cracks are automatically identified, and the geometric characteristics of the diseases are calculated. The collection speed of the tunnel rapid detection vehicle based on this technology can reach 60–80 km/h [10, 11], which is significantly higher than the equipment based on laser scanning technology. It is suitable for rapid detection of apparent diseases of high-grade highway tunnels with high minimum speed limit and large traffic flow under the condition of no traffic control.

A rapid tunnel inspection equipment using infrared supplemental light and its tunnel crack measurement technology based on digital images are introduced in this paper. According to relevant specifications, the crack width of reinforced concrete structure of tunnel lining shall not exceed 0.2 mm, and the crack width of concrete structure shall not exceed 0.4 mm [12]. In order to ensure that the crack detection accuracy of tunnel rapid detection vehicle can meet the specification requirements, scientific and reasonable calibration methods shall be adopted to verify it. In view of the shortcomings of existing conventional calibration methods, a new calibration method for crack detection accuracy is proposed in this paper. This method can effectively solve the problems of the current tunnel rapid detection vehicle crack detection accuracy calibration method, such as inconsistent with the actual environment of the operating tunnel, poor accuracy, affecting the normal operation of the tunnel, heavy workload of image processing, low efficiency of extracting target crack information. The reliability of the calibration method is verified by the real tunnel operation.

2 Tunnel Image Acquisition Mode Based on Infrared Supplement Light

At present, with the exception of some very short-distance tunnels, the long and extra-long tunnels in my country's highway tunnels that have been opened to traffic account for a relatively large [13]. The light inside such tunnels is relatively dim, and lighting systems are usually installed in the tunnels. In addition, there are lights of driving vehicles in the tunnel. Together with the illumination of the tunnel lighting system, the brightness in the tunnel will be uneven and halo is prone to occur. These will cause great interference to the imaging quality of the camera, and even lead to collection. The image is blurred and cannot be used for image analysis of lining cracks.

The existing supplementary light sources mainly include laser light sources, white light sources and infrared light sources. For the needs of rapid detection in tunnels, high-intensity white light lamps will instantly blind drivers in the surrounding lanes, which will seriously affect traffic safety. Although laser light sources have many advantages, they are harmful to human eyes and will cause blindness in severe cases. In view of the defects in the use of white light source and laser light source in the tunnel, Shanghai Tongyan rapid inspection vehicle adopts the infrared supplement light mode. Because infrared light is invisible light, it will not affect the safety of surrounding traffic. The infrared band is enhanced to achieve better imaging results.

3 Tunnel Lining Crack Measurement Technology Based on Digital Images

Photographic measurement technology based on digital image processing has penetrated into the fields of civil engineering such as crack width measurement, deformation monitoring, and rock mass identification [14–16]. Based on image processing technologies such as optimal threshold segmentation, morphological processing, edge extraction, and sub-pixel methods, this section proposes an automatic recognition algorithm for tunnel lining cracks, which is different from the traditional short-range image processing defects in the image processing of long-distance imaging pictures. Hand-held contact type crack width measurement method, using ordinary digital camera to collect pictures from long-distance digital photography, and quantitatively analyze the crack width on the lining surface.

3.1 Crack Area Extraction

Threshold segmentation is performed on the preprocessed image. Threshold segmentation is the most common technique for direct detection of regions. However, in the whole image, the gray scale has a large variation range and a single threshold cannot meet the requirements. An adaptive threshold related to the gray characteristics of the region is dynamically adopted through an improved loop iterative algorithm. The segmentation of the crack area is mainly carried out through three steps: the first step is area selection, the second step is threshold segmentation, and the third step is morphological processing.

The first step is to interactively select the area you want to process through the mouse. You can select a smaller area and perform image processing on it to prevent processing deviations due to the large picture and small target cracks, and turn the selected area into gray Degree image.

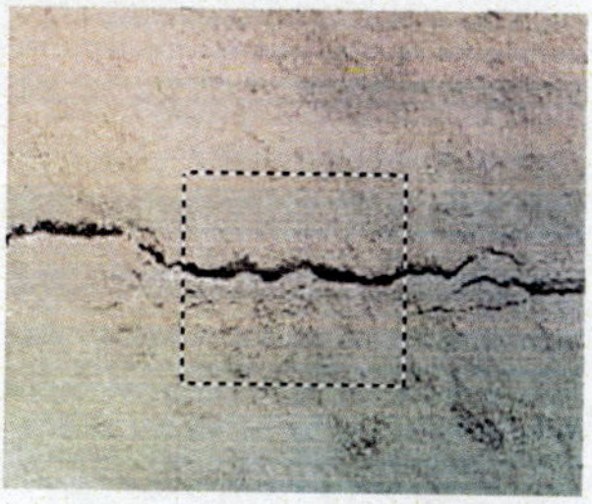

Fig. 1. Regional selection

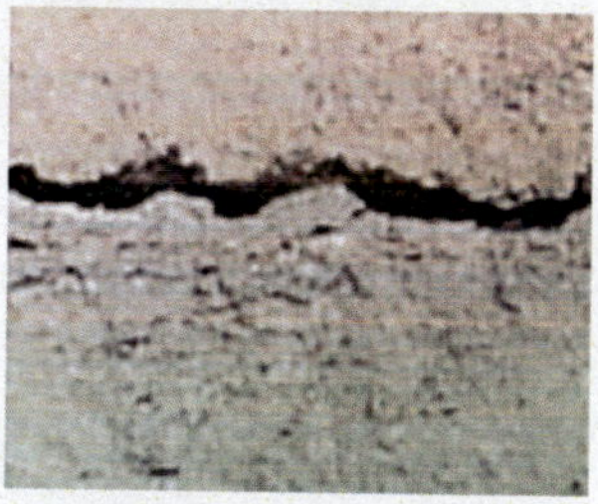

Fig. 2. Magnified view of selected area (grayscale image)

The second step is to perform threshold segmentation on the image of the selected area, and use an iterative algorithm to adaptively obtain the global threshold. Assuming that the crack area is the darkest area in the entire image, the specific steps of the iterative algorithm are as follows:

(1) Calculate the average gray valuc T_{ave} of the entire image, the calculation formula is as follows:

$$T_{ave} = \sum Pixels \Big/ N \tag{1}$$

Among them, $\sum Pixels$ represents the sum of gray levels of each pixel in the image, and N represents the total number of pixels. Set T_{ave} as the initial threshold T_k, at this time $k = 1$.

(2) According to T_k, the image is divided into two parts, the target and the background, the area with a gray level less than T_k is called the target area, and the area greater than T_k is called the background area, and then the average gray value of the two areas is calculated and denoted as T_{low} and T_{high}.

(3) Using T_{low} and T_{high} to find a new threshold, the calculation formula is as follows:

$$T_{k+1} = \left(T_{low} + T_{high}\right)/2 \tag{2}$$

If T_{k+1} is not equal to T_k, then set $T_k = T_{k+1}$, return to step (2), and set $k = k + 1$; at the same time; if T_{k+1} is equal to T_k, then go to step (4);

(4) Using T_{k+1} as the final segmentation threshold for binary image segmentation, the target area basically contains the crack area, and there are also some noise points and some small areas with gray values that are not much different from the cracks. The image is black and white. Reverse, as shown in Fig. 3.

Fig. 3. Threshold segmentation

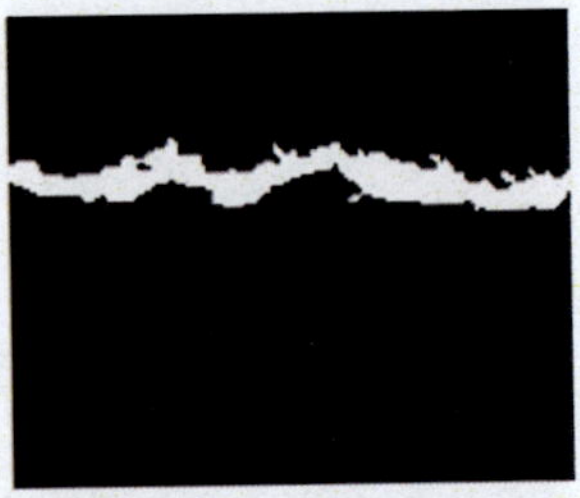

Fig. 4. Maximum target area

After the above segmentation steps, find the largest part of the image target area (that is, the largest white part), and assign 0 (black) to other areas to extract the area containing cracks, but it may also contain large blocks at the same time Background area, so further judgment is required.

Perform an etching operation on the n*n square unit matrix and the image (n is an integer greater than zero, the crack width in the above picture does not exceed 15 pixels, so take n = 15) to see if there is a matching area.

If there is, it means that in addition to the crack area, there is a large background area divided into the target area in the extracted area, and the following processing is performed:

Since the crack area is the darkest area in the entire picture, we can use the local threshold method to average the gray levels of all points in the original image corresponding to the extracted area. The calculation formula is as follows:

$$T_j = \sum Pixels_{\mathrm{t\,arg\,et}} \Big/ N_{\mathrm{target}} \tag{3}$$

Divide the image into two parts of the target and the background according to T_j. Repeat step (5) after the black and white inversion of the image until there is no area matching the unit matrix in the image, indicating that there are no large non-narrow and long areas in the target area and cracks have taken up the majority of the target area. At this time, the extraction effect is good, as shown in Fig. 4.

The third step is to use morphological methods to open and close the image, fill holes, remove edge burrs, etc., and then find the largest white area again, and assign the remaining areas to 0 to ensure that only the cracks in the image are white areas, and finally as shown in Fig. 5 and Fig. 6.

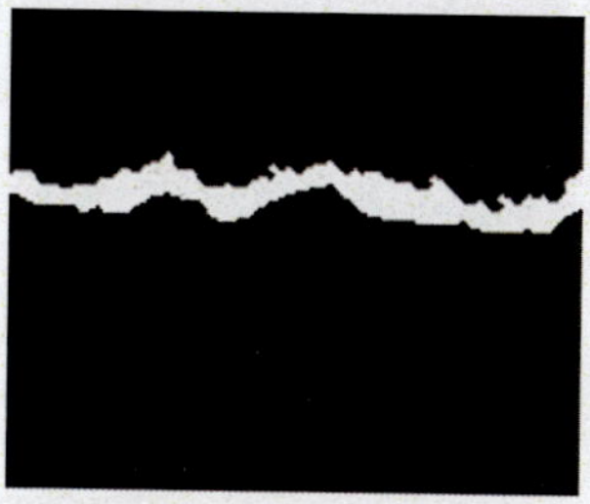

Fig. 5. Filling holes and removing edge burrs

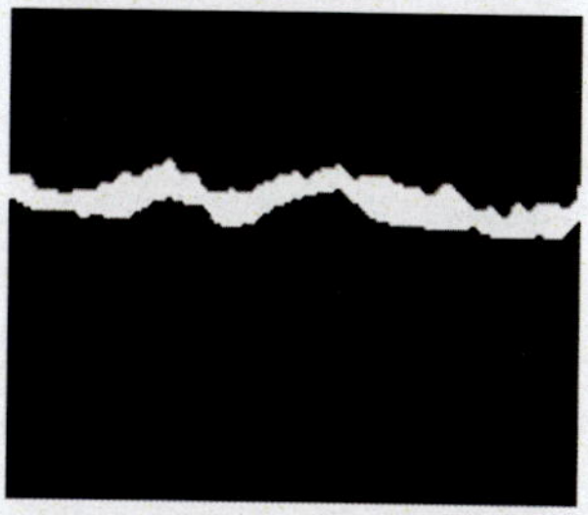

Fig. 6. Smoothing the edges further

3.2 Crack Edge Extraction Based on Zernike Orthogonal Moments

The most basic feature of cracks is the edge. Edge extraction is the most important core step in the whole process. The quality of the edge detection algorithm will greatly affect the detection effect and accuracy. In order to improve the accuracy of crack edge extraction, the algorithm uses The image sub-pixel edge detection algorithm based on Zernike orthogonal moments is an ideal step edge model.

In Fig. 7, k is the step height, h is the background gray level, l is the vertical distance from the center of the disc to the edge, and ϕ is the angle between the edge and the y axis.

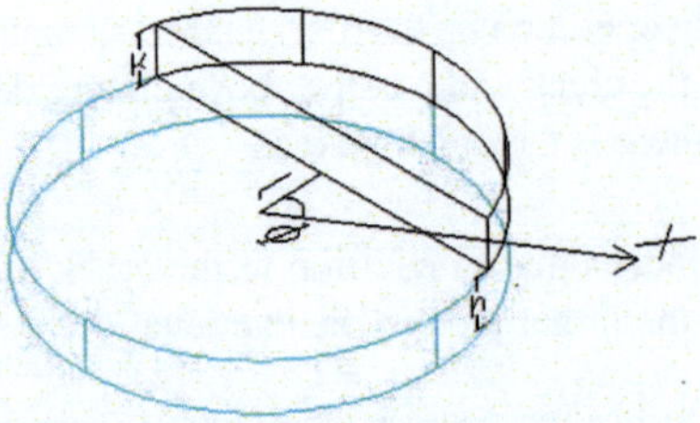

Fig. 7. Ideal step edge model

First obtain the edge parameters of each pixel, then perform post-processing on the *k* value to determine the threshold, and then perform edge refinement after binarization to obtain the final edge image, as shown in Fig. 8 and Fig. 9.

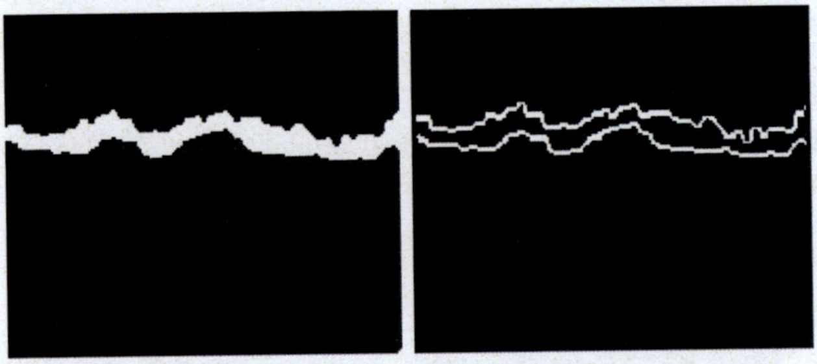

(a) Original image (b) Edge extraction image

Fig. 8. Sub-pixel edge detection

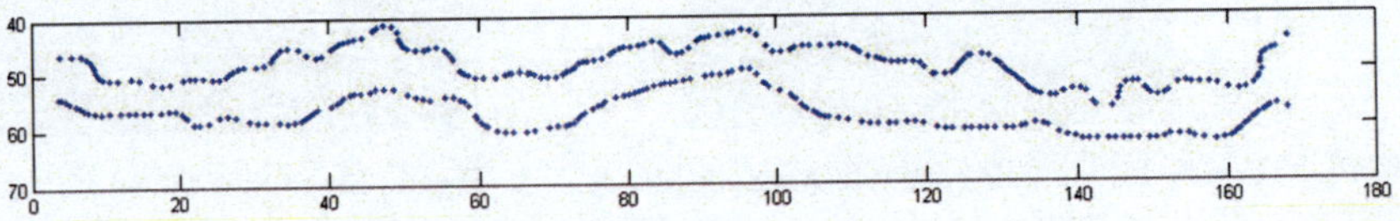

Fig. 9. Precise edge map composed of sub-pixels

3.3 Calculation of Crack Width

Distinguish the upper and lower edges of the crack, select each point on the upper edge respectively, and use the "minimum distance method" to calculate the width of the target crack.

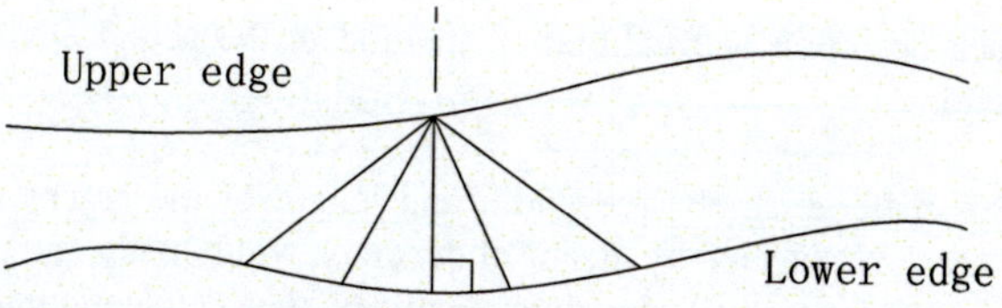

Fig. 10. Schematic diagram of the minimum distance method

As shown in Fig. 10, according to the coordinates of the upper and lower edge points in the vertical direction, start from the first point of the upper edge, and use the coordinates of the upper edge point to respectively and each coordinate point of the lower edge using the distance formula between two points in advanced mathematics Calculate, the minimum value calculated as the distance from the point to the lower edge can be expressed as:

$$w_i = \min\left(\sqrt{(x_i - x_k)^2 + (y_i - y_k)^2}\right) \tag{4}$$

Where $k = 0, 1, 2, 3, \cdots$, the formula means to take the minimum value of the distance from the point i of the upper edge to all the points on the lower edge, and calculate the distance value from each point of the upper edge to the lower edge in turn, and combine these distances The average value or maximum value is calculated, that is, the average value and maximum value of the crack width are recorded as $\bar{w}$ and $w_{\max}$ respectively. The calculation formula is as follows:

$$\bar{w} = \sum_i w_i \Big/ n \tag{5}$$

$$w_{\max} = \max(w_i) \tag{6}$$

4 Conventional Calibration Method

At present, there are two methods to calibrate the crack detection accuracy of tunnel rapid detection vehicle.

(1) Calibration method of model crack installed on Tunnel Model Bench.
In this method, a small full-scale tunnel crack model (generally 0.5 m–1 m in length) is built, and the crack image is collected statically by the tunnel fast detection vehicle. After the post-processing analysis, the size data of the model crack is obtained, and then compared with the actual size data of the model crack, so as to achieve the purpose of crack detection accuracy calibration.
Due to the large difference between the tunnel model in this method and the actual internal environment of the highway tunnel, the influence of the tunnel irregular section profile, illumination, vehicle exhaust and dust, vehicle running speed, etc. is not considered comprehensively, resulting in the lack of applicability and accuracy of the calibration results to the actual detection environment.

(2) Crack detection and calibration method in the actual environment of operating tunnel.

In this method, firstly, the width and length of the cracks on the surface of the tunnel lining are detected by means of the crack width meter and the tape. Secondly, the image of the cracks is collected dynamically by the vehicle mounted instrument, and the corresponding cracks and their size information are identified by the image processing in the later stage. Finally, the cracks collected manually are compared with the cracks collected by the tunnel rapid detection vehicle. The crack information is processed to achieve the purpose of calibrating the crack detection accuracy.

Because every time before collecting the crack size of the arch or side wall of the tunnel manually, traffic control is needed, and then the crack data can be collected with the help of the aerial vehicle, which not only affects the normal operation of the tunnel, but also costs more manpower. In addition, all kinds of cracks in the tunnel (including circumferential cracks, longitudinal cracks, oblique cracks, cracks, etc.) are mostly randomly distributed in different locations of different mileage in the tunnel, which results in a large amount of image data to be collected by the fast detection vehicle of the tunnel, and a heavy work of image processing in the later stage, which is not conducive to the rapid extraction of the crack information to be compared in the collected images.

5 New Methods

In order to reduce the amount of target image data that need to be collected by the tunnel rapid detection vehicle, the cracks of various types and sizes are concentrated and randomly arranged on the model plate in advance. In the later stage, only the images of the model plate area need to be intercepted for crack recognition, extraction and size calculation, which greatly improves the work efficiency. Because the model plate is installed in the operation highway tunnel, that is, the calibration environment comprehensively considers the influence of tunnel section profile, illumination conditions, vehicle exhaust, dust and other factors on the detection accuracy of vehicle crack in the rapid detection of the tunnel, which meets the application requirements of the actual project.

5.1 Implementation Steps

The implementation steps of the method mainly include:

(1) The cracks of different types and sizes are drawn on the rectangular model plate with similar color to the inner surface of the tunnel lining, and the model plate with simulated cracks is installed on different parts of the inner surface of the tunnel lining, including the side wall and arch of the tunnel.
(2) The crack image on the model plate is dynamically collected by the tunnel fast detection vehicle, and the size information of the crack is obtained through the image processing and analysis in the later stage.

(3) By comparing the crack information obtained by the tunnel rapid detection vehicle with the actual size of the cracks on the model plate, the accuracy of the crack detection of the tunnel rapid detection vehicle can be calibrated.

5.2 Field Test

5.2.1 Production and Installation of Model Cracks

1) Technical performance requirements of crack plate.
The technical performance of the crack slab mainly includes the background color difference, crack color, crack slab and the dimensional accuracy requirements of the crack slab. The background color of the crack slab should be as close as possible to the surface color of the tunnel secondary lining, and the specific requirements are as follows. ① When the crack plate is installed on the side wall of the tunnel with white tiles, the background color difference of the crack plate ΔE (The unit is NBS) meets $\Delta E \subset [0.5,\ 1.0]$. When the crack plate is installed on the concrete surface of the tunnel arch, the background color difference of the crack plate ΔE meets $\Delta E \subset [0.5,\ 1.5]$. Because the color of the actual crack is darker than that of the concrete background, this method chooses black as the color of the simulated crack.

$$\Delta E = \sqrt{(\Delta L^2 + \Delta a^2 + \Delta b^2)} \tag{7}$$

Where ΔL is the lightness difference, Δa is the chromaticity difference, and Δb is the hue difference.
② The thickness of the crack plate shall not be greater than 5 mm, so as to avoid that the crack plate is too thick and protrudes on the surface of the second lining of the tunnel, which will affect the accuracy of image acquisition.
③ The cracks of different types, lengths and widths are drawn in a certain frame range by CAD software, and the background color and crack color of the frame area are set up. The printing error of the crack width is not more than 5%, and the printing error of the crack length is not more than 1%, so as to ensure that the dimensional accuracy of the cracks on the crack plate meets the requirements.
2) Crack plate installation requirements.

The installation of crack plate needs to consider the tunnel outline size, installation position, illumination environment and other factors. The specific requirements are as follows.

① According to the requirements of tunnel outline size, crack plates should be installed in two lane (see Fig. 11) and three lane (see Fig. 12) tunnels respectively;
② The installation position includes the side wall and arch of the tunnel;
③ Four sections of tunnel entrance section, transition section, middle section and exit section are selected, corresponding to different illuminance environment in the tunnel. Dark area with low illuminance, moderate illuminance area and high illuminance area around the lighting lamps are selected for each section as far as possible to install crack plate;

④ It is applicable to the tunnel with emergency parking zone. The installation section of crack plate is divided into normal section and widened section.

5.2.2 On-site Inspection Steps

① The average detection speed of the fast detection vehicle of the tunnel shall not be less than 60 km/h. Using stopwatch timing to obtain the single detection time t of detection vehicle, according to the tunnel design data to obtain the single detection mileage l of detection vehicle, the average speed is $v = l/t$.

② According to the configuration requirements of different vehicle detection systems, the driving lanes of vehicles are selected independently. In the case of one-way two lane tunnel, the fast detection vehicle drives in the right main lane and the left overtaking lane for two times. In the case of one-way three lanes, the fast detection vehicle of the tunnel is required to drive in the right most slow lane and the left most fast lane for two times of detection (to avoid the occlusion effect of passing vehicles in the adjacent lane on the target detection object).

5.3 Calibration Index

The calibration indexes of the crack detection accuracy of the tunnel rapid detection vehicle include the following 6 items.

① Crack identification rate: Count the number of simulated cracks identified by the fast detection vehicle of the tunnel. Crack identification rate = the number of simulated cracks identified / the total number of simulated cracks. ② Measurement error of crack length: Calculate the difference between the crack length identified by the rapid detection vehicle of the tunnel and the corresponding simulated crack length. Measurement error of crack length = | identified crack length - corresponding simulated crack length |. ③ Measurement error of crack width: Calculate the difference between the crack width identified by the rapid detection vehicle of the tunnel and the corresponding simulated crack width. Measurement error of crack width = | identified crack width - corresponding simulated crack width |. ④ Accuracy rate of crack type discrimination: Judge whether the type of crack identified by the rapid detection vehicle of the tunnel is consistent with the corresponding simulated crack type, and the accuracy rate of crack type discrimination = the number of correct cracks identified by the crack type discrimination / the total number of simulated cracks identified. ⑤ Stake number error: Calculate the difference between the stake number of the crack identified by the rapid detection vehicle of the tunnel and the actual stake number of the simulated crack. Stake number error = | the stake number of the identified crack - the actual stake number of the simulated crack. ⑥ Generation and derivation of crack distribution map: it is required that the crack distribution map is clear, and the human eye can clearly identify the crack and background image on the distribution map.

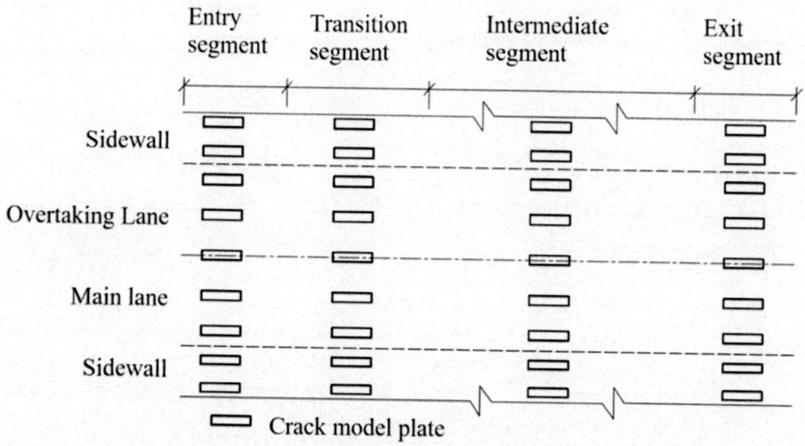

(a) Layout plan

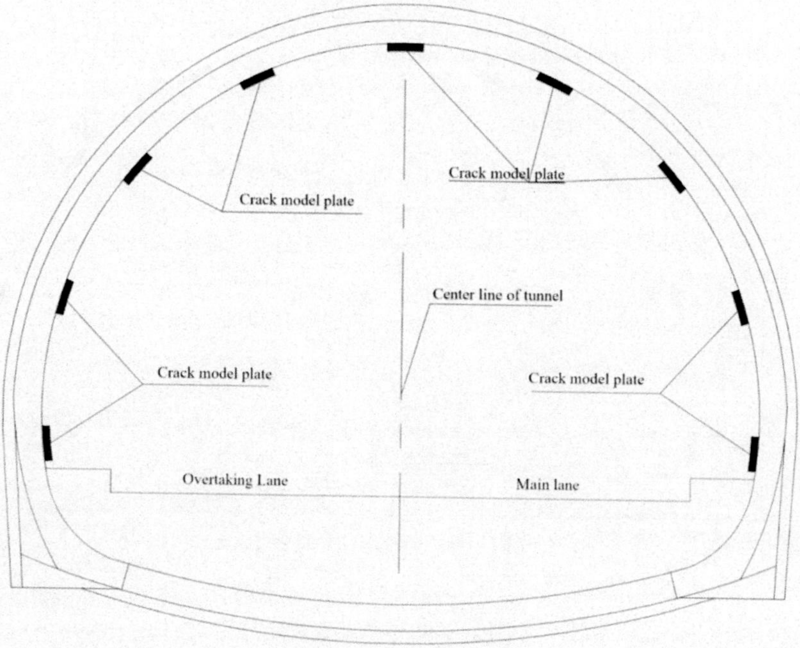

(b) Cross section layout

Fig. 11. Installation diagram of crack plate of two lane tunnel

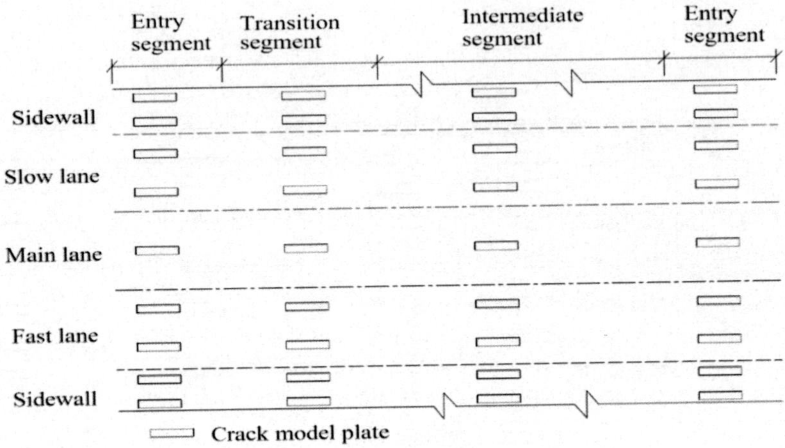

(a) Layout plan

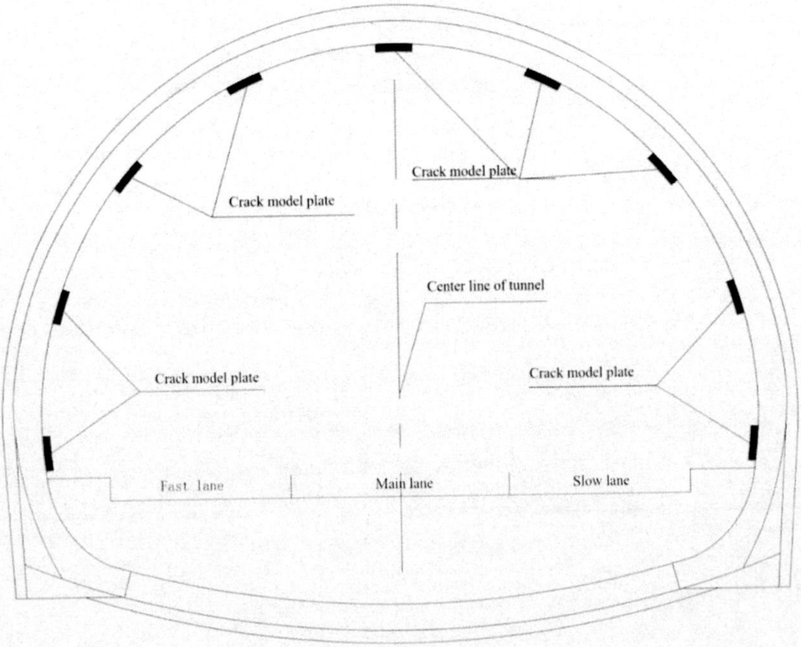

(b) Cross section layout

Fig. 12. Installation diagram of crack plate of three lane tunnel

5.4 Qualification Standard

The qualification criteria of crack detection accuracy calibration index are as follows. ①The crack recognition rate of width not less than 0.2 mm reaches 100%. ② The measurement error of crack length shall not be greater than 100 mm. ③ The measurement error of crack width shall not be greater than 0.1 mm. ④ The accuracy of crack type identification shall not be less than 90%. ⑤ The stake number error shall not be greater than 1 m.

6 Case Studies

Two lane highway tunnel T1 and three lane highway tunnel T2 in Guangdong Province of China are selected. The calibration method proposed in this paper is used to verify the crack detection accuracy of three kinds of conventional fast detection vehicles (respectively recorded as vehicle a, vehicle B and vehicle C). Before the test, mileage stake number plate and simulated crack plate shall be set on the side wall of the tunnel in advance, and then the rapid detection vehicle of the tunnel shall be used for data collection; after the field test, the data collected by the rapid detection vehicle shall be processed and compared with the simulated crack information set manually, so as to obtain the calibration index of crack identification rate, size identification accuracy, etc. of the rapid detection vehicle of the tunnel Mark, and judge the calibration result according to the qualified standard.

6.1 Implementation Plan

(1) Setting of mileage station.
Mileage marks shall be made along the inner side wall of the inspection tunnel, and a mileage sign shall be set every 10 m.

(2) Production and installation of crack plate.
The pre drawn simulated cracks are shown in Fig. 13. In this test, 4 crack plates are installed on the right line of T1 tunnel (two sections of portal section and middle section are installed on the side walls on both sides), and 6 crack plates are installed on the left line of T2 tunnel (two sections of portal section and middle section are installed on the haunches of the side walls and middle section on both sides). The installation position of crack plates on the side walls is shown in Fig. 14.

(3) Data collection and analysis.
① Collect and analyze the data of the fast detection vehicle of the tunnel. The rapid vehicle is used to collect the data of the target test tunnel to obtain the information of the pile number, location, type and size of the apparent crack disease of the tunnel lining. ② This paper classifies and counts the simulated crack diseases on the artificial crack plate in detail. ③ Based on the data of artificial crack simulation as the comparison standard, verify whether the crack detection results of the rapid detection vehicle completely cover the artificial crack simulation, as well as the accuracy of the pile number location, type judgment and size measurement of the crack, and analyze the display effect of its lining scanning image.

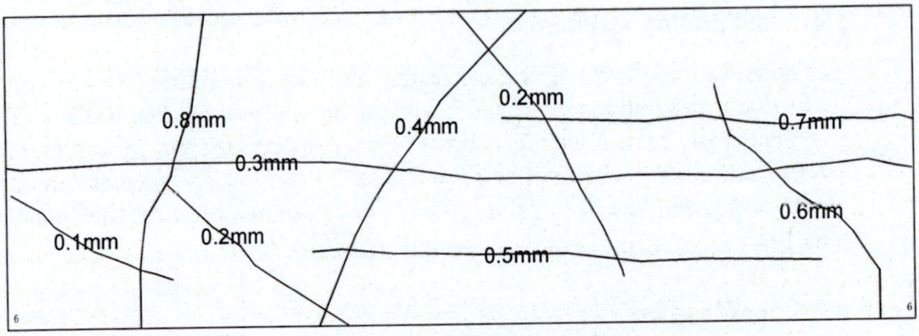

Fig. 13. Schematic diagram of simulated crack plate and crack drawn

Fig. 14. installation position of crack plate on side wall of three lane tunnel.

6.2 Test Results

6.2.1 T1 Tunnel

(1) Calibration index verification results.
According to Table 1 and Table 2, for cracks with a width of not less than 0.2 mm, the recognition rate of the three detection vehicles is above 95%, which is consistent with the crack detection accuracy of the detection vehicle identification. Except for vehicle A, the other two detection vehicles meet the qualification standard specified in this calibration method (i.e. the recognition rate is 100%), and most of the measurement results with a crack width of 0.1 mm are still missing. There is a large error, and the other indicators meet the qualification standard.

(2) Layout effect of crack board.
It can be seen from Fig. 15 that the scanned images of car A and car C are relatively clear, car A is spliced continuously, and the visual effect is better. The splicing seams

Table 1. Calibration indexes of crack detection accuracy of different rapid detection vehicles (T1 tunnel)

Detection vehicle	Crack identification rate /%		Average measurement error of crack length/mm	Average measurement error of crack width/mm	Accuracy rate of crack type identification /%	Pile mileage error/m
	Width ≥ 0.2 mm	Other width				
A	96.8	88.6	60	0.18	91.4	<1
B	100	96.2	40	0.42	97.8	<1
C	100	100	40	0.13	94.0	<1

Table 2. Cracks missed by different rapid detection vehicles (T1 tunnel).

Undetected cracks	A vehicle		B vehicle	C vehicle
Width/mm	0.1	0.2	0.1	No
Quantity/number	3	1	2	No

of car B and car C are obvious, the image staggered seams are not continuous, and the display effect is poor.

6.2.2 T2 Tunnel

According to Table 3 and Table 4, for cracks with a width of not less than 0.2 mm, except for car A, the recognition rate of the other two detection vehicles is above 95%. The result is consistent with the crack detection accuracy of the detection vehicle identification, but only car C meets the qualification standard specified in this calibration method, and most of the missed crack width is 0.1 mm. The measurement results of crack width only car C meets the qualification standard. The other indexes meet the qualification standard.

The calibration method proposed in this paper is used to calibrate the crack detection accuracy of the same fast detection vehicle in different highway tunnels. It is found that the results of the same calibration index do not necessarily meet the qualification standard. The analysis may be related to the accuracy calibration of crack detection for different types of tunnels and internal environment by the tunnel rapid detection vehicle, and the inconsistent setting of hardware acquisition parameters of the detection vehicle.

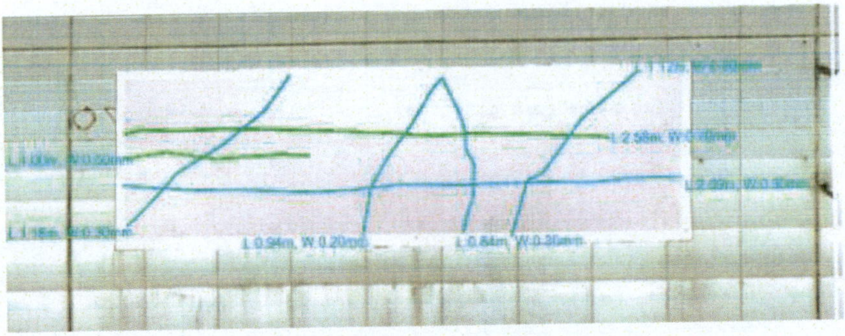

(a) A car

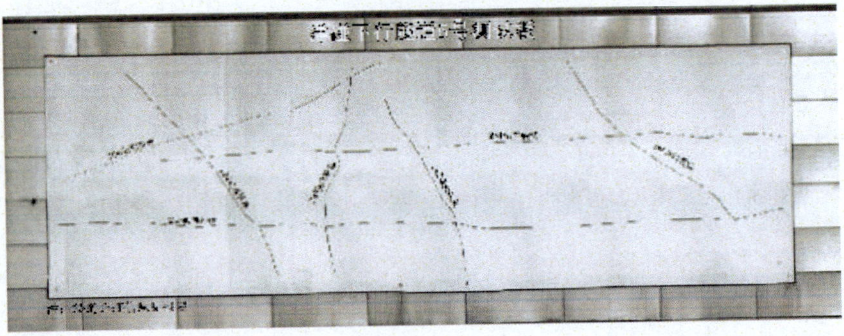

(b) B car

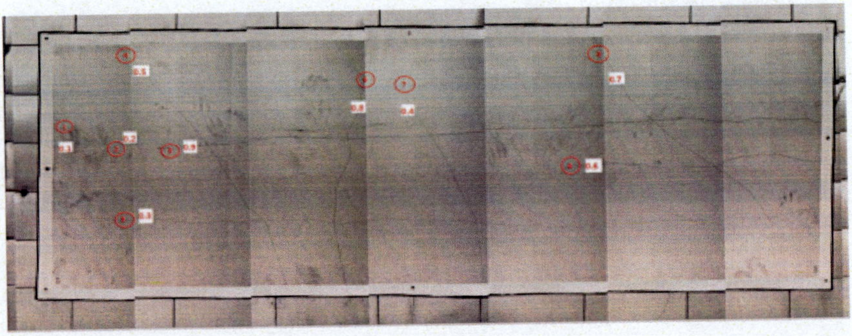

(c) C car

Fig. 15. Scanning image of crack plate of inspection vehicle

Table 3. Calibration indexes of crack detection accuracy of different fast detection vehicles (T2 tunnel)

Detection vehicle	Crack identification rate /%		Average measurement error of crack length / mm	Average measurement error of crack width / mm	Accuracy rate of crack type identification /%	Pile mileage error / m
	Width ≥ 0.2mm	Other width				
A	93.1	81.8	50	0.22	95.2	<1
B	95.4	87.5	60	0.16	95.2	<1
C	100	97	60	0.06	100	<1

Table 4. Cracks missed by different rapid detection vehicles (T2 tunnel).

Undetected cracks	A vehicle		B vehicle		C vehicle
Width /mm Quantity / number	0.1	0.2	0.1	0.2	0.1
	4	2	3	1	1

7 Conclusions

(1) The calibration method proposed in this paper makes up for the shortcomings of the two conventional calibration methods, that is, it conforms to the authenticity of the calibration environment and the operating tunnel environment, and can improve the efficiency and precision of the accuracy calibration of the tunnel rapid vehicle crack detection in the tunnel. It is also applicable to the calibration of crack width detection accuracy for tunnel rapid detection vehicle with infrared or visible light illumination.

(2) The application case study shows that for cracks with width no less than 0.2 mm, the crack identification rate, average measurement error of crack length, accuracy rate of type discrimination and pile number error of the rapid detection vehicle can meet the qualification standards specified in this method, and the vast majority of the missed cracks are those with width less than 0.2 mm.

(3) It is verified that the crack width detection accuracy of the vehicle identification is 0.2 mm. For the larger average measurement error of crack width, the accuracy of calibration results can be verified by repeated calibration.

(4) After a large number of verification of the actual fixed inspection of the tunnel, when the inspection speed is 80 km/h, there is no need for vehicles to follow or road escort, the crack width identification accuracy is 0.2 mm, the full width of two lane tunnel is collected once, and the full width of three and four lane tunnel is collected twice.

(5) The calibration method proposed in this paper can be used to calibrate the tunnel crack disease information and realize continuous, dynamic and complete collection and analysis. It should be noted that the width of each crack in this test is the same value, and the crack background is single and uniform, but in fact the width of the same crack

is often different, and the background is rough. For the former, the maximum or average crack width can be used in actual detection, while the latter can be solved by improving the image resolution or improving the algorithm of crack edge extraction.

Acknowledgements. The authors gratefully acknowledge the funding from the Science and technology plan project of Guangdong Provincial Communications Department (2016-02-015). The authors thank anonymous reviewers for their insightful comments and suggestions that have helped to improve the clarity and quality of this manuscript.

References

1. Sansalone, M., Carino, N.J.: Detecting delaminations in concrete slabs with and without overlays using the impact-echo method. Mater. J. **86**(2), 175–184 (1989)
2. Davis, A.G., Lim, M.K., Petersen, C.G.: Rapid and economical evaluation of concrete tunnel linings with impulse response and impulse radar non-destructive methods. NDT E. Int. **38**(3), 181–186 (2005)
3. Rongben, W., Chao, W., Xiumin, C.: Developments of research on road pavement surface distress image recognition. J. Jilin Univ. Eng. Technol. Edn. **32**(4), 91–97 (2002)
4. Yu, S.N., Jang, J.H., Han, C.S.: Auto inspection system using a mobile robot for detecting concrete cracks in a tunnel. Autom. Constr. **16**(3), 255–261 (2007)
5. Lee, B.Y., Kim, Y.Y., Yi, S.T., et al.: Automated image processing technique for detecting and analysing concrete surface cracks. Struct. Infrastruct. Eng. **9**(6), 567–577 (2013)
6. Pingrang, W.: Analysis and comparison of current situation of rapid detection technology for tunnel diseases. Highways Autom. Appl. **3**, 241–245 (2016)
7. Shen, Z.Y., Tan, Z.: Application research on GRP5000 tunnel inspection car in Shanghai subway. Shanxi Archit. **39**(27), 158–159 (2013)
8. Fangqi, W., Yonghua, L., Qiuzhuo, L.: Application of non-contact 3D infrared scanning vehicle system in tunnel regular detection. Highway Tunnel **1**, 40–44 (2016)
9. Yuan, Y.: System for quickly detecting tunnel deformation. U.S. Patent Application 15/787,259 (2018)
10. Huang, H., Sun, Y., Xue, Y., et al.: Inspection equipment study for subway tunnel defects by grey-scale image processing. Adv. Eng. Inform. **32**, 188–201 (2017)
11. Yang, R.: Study on Impact of Mountain Freeway Tunnel Entrance Environment on Drivers' Physiology and Behavior. South China University of Technology (2017)
12. Zongfan, L.: Direct control of allowable crack width of reinforced concrete flexural members. Ind. Construct. **11**, 43–55 (1988)
13. Jiang, S., Lin, Z., Wang, S.: China's road tunnel development in 2018. Tunnel Construct. **039**(007), 1217–1220 (2019)
14. Zhang, J., Sha, A., Gao, H., et al.: Cracks automatic identification and evaluation system based on digital image processing pavement. J. Chang'an Univ. Nat. Sci. Edn. **024**(002), 18–22 (2004)
15. Liu, Q., Tan, Q.: Concrete crack detection based on image processing. J. Wuhan Univ. Technol. **04**, 69–71
16. Lan, Y.: Measurement and analysis of crack feature of concrete surface based on digital image processing techniques. Southeast University (2006)

Research on the Construction of 3D Laser Scanning Tunnel Point Cloud Based on B-spline Interpolation

Zheng Wei[1,2](✉), Tianyu Yao[3], and Chenghui Shi[2]

[1] Southest University, Nanjing 211189, China
[2] Traffic Engineering Management Center of Zhejiang Province, Hangzhou 310005, China
[3] Zhejiang University of Technology, Hangzhou 310014, China

Abstract. In this paper, the three-dimensional reconstruction of the tunnel point cloud is carried out based on the polynomial fitting method. Due to some shortcomings, a method of constructing a curved surface by region is introduced to analyze its fitting accuracy and fitting effect. For the tunnel vault with larger (or large) curvature, a B-spline interpolation method is used to reconstruct the tunnel, and for the arch waist, a polynomial fitting method is used to optimize the fitting method in different regions. Finally, the application of fitting results is prospected on this basis.

Keywords: Railway tunnel · B-spline · Polynomial fitting

At present, the rail transit industry of China is developing rapidly, with 140000 km of existing operating railways, and its achievements have attracted worldwide attention. With the rapid development of railway construction, the requirements for the thickness control of secondary lining in NATM tunnel construction have been further improved. At the same time, railway tunnel safety accidents have occurred frequently in recent years, so it is very important to monitor the progress of tunnel construction more effectively and timely at this stage.

Point cloud 3D reconstruction is widely used in various areas, not only in building structure [1], but also in the area of surveying and mapping [2]. However, there is still less research in the area of tunnel, and more in-depth exploration is required. Curved surface reconstruction usually adopts parametric curved surface reconstruction and mesh reconstruction, etc. [3]. In tunnel engineering, 3D laser scanning technology is gradually introduced for data collection, which can cover the tunnel excavation surface in a high-speed and all-round way and acquire tunnel point cloud data [4], laying a foundation for the application of tunnel point cloud data. However, point cloud data is scattered data, and there are obstacles to its direct application, so it is necessary to pretreatment point cloud before application. This paper aims to adopt the parametric curved surface reconstruction method to reconstruct the tunnel curved surface.

J. J. Yang et al. (Eds.): GeoChina 2021, SUCI, pp. 111–118, 2021.
https://doi.org/10.1007/978-3-030-79672-3_8

1 Point Cloud Data Processing

In this paper, a Faro x330 3D laser scanner is used to collect data from Leicao Mountain Tunnel in Lishui City, Zhejiang Province as shown in Fig. 1.

Fig. 1. Collected data on the spot

After the collection, three steps of point cloud positioning and registration, denoising and thinning are used to process the data. The specific steps are as follows:

(1) The station position is calibrated by the absolute target positioning method [5]. Each station is spatially located by the corresponding target ball. In the long tunnel section, positioning efficiency and accuracy can be improved.

(2) In the process of data acquisition, because of point clouds that are not related to the tunnel due to construction equipment and noise, the redundant points and discrete points of tunnel point clouds are eliminated by the Gaussian filtering method [6].

(3) Due to the large amount of point cloud data acquired, in order to improve the calculation efficiency, the equal interval method is used to dilute the point cloud data, so that the point cloud is arranged according to a certain interval.

The three-step process of point cloud data processing is shown in Fig. 2.

2 Tunnel Surface Construction

In this paper, taking a single-track railway tunnel as an example, excavation is carried out by NATM [7], and the design surface is shown in Fig. 3.

As seen from Fig. 3, the design surface of the tunnel is composed of five arcs, each with its center, resulting in the curvature of the tunnel vault being far greater than other positions. Therefore, this paper proposes to construct a tunnel surface based on the tunnel point cloud. According to different curvatures, it can be divided into three regions: 0–1/4 π, 1/4 π–3/4 π, 3/4 π–π. They are divided into two regions: vault and haunch.

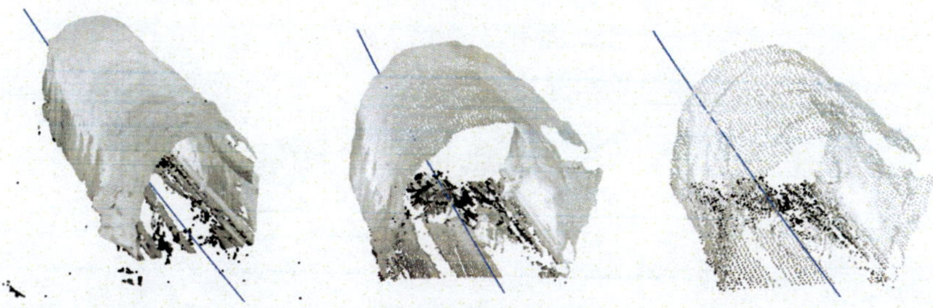

(a) Original tunnel point cloud (b) Tunnel point cloud after denoising (c) Tunnel point cloud after thinning

Fig. 2. Schematic diagram of point cloud data processing process

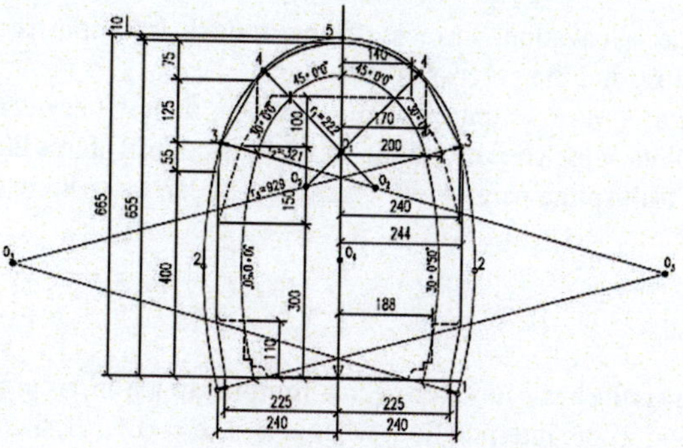

Fig. 3. Design surface of tunnel

2.1 Construction of All Area Polynomial Surface

Firstly, polynomial fitting is carried out for the tunnel surface in the whole area. The design central axis is used as the x-direction of the bottom surface. There will be repetition points in the z-direction. The left and right sides of the arch waist need to be calculated separately, and the results are shown in Fig. 4.

According to Fig. 4, the fitting surface can construct the tunnel surface smoothly and completely, and the fitting effect at the arch waist is closer to the real surface, while the vault area is partially distorted.

2.2 Construction of Subdivision Surface

Based on the whole area fitting results, the method can restore the tunnel surface better at the arch waist with a smaller curvature, but there will be some distortion at the vault with a larger curvature. In order to better reconstruct the three-dimensional surface of the tunnel, this paper uses the sub-regional tunnel reconstruction.

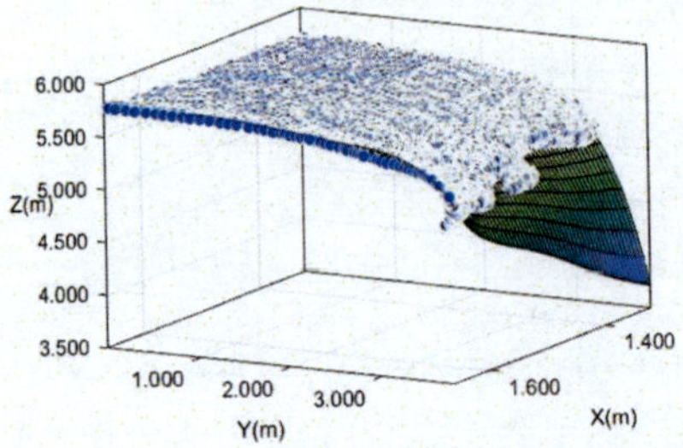

(a) Left side of fitting face

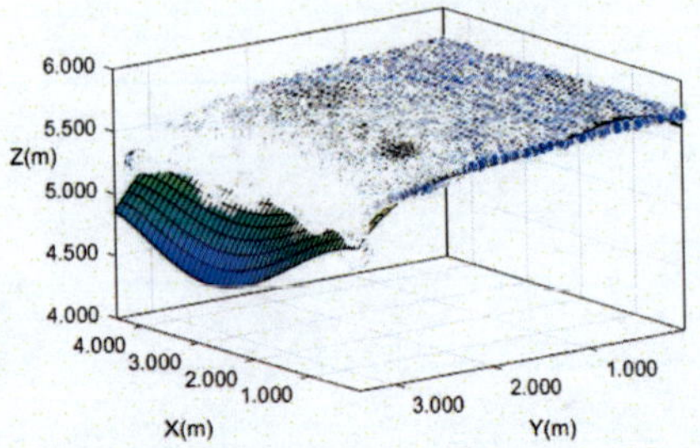

(b) Right side of fitting face

Fig. 4. Full area fit face

2.2.1 Construction of Vault Surface

Firstly, the excavation surface of the tunnel is reconstructed by cubic B-spline interpolation, and the process is as follows:

Given $m + n + 1$ control points in space, then the surface can be solved by these control points, which represents the set of these control points like $P_h(h = 0, 1, 2, ..., m + n)$, so the n-th spline parameter expression [8] can be written:

$$P_{i,n}(t) = \sum_{k=0}^{n} P_{i+k} G_{k,n}(t),\ t \in [0, 1] \tag{1}$$

$G_{k,n}(t)$ is the basic function of an n-th spline function, t represents the node parameter.

In cubic spline interpolation, when $n = 3$, $k = 0,1,2,3$, there will be 4 control points P_0, P_1, P_2, P_3 in the control polygon to represent the surface. At this time, the cubic spline curve is a cubic polynomial [9].

$$G_{0,3}(t) = \frac{1}{6}(-t^3 + 3t^2 - 3t + 1) \tag{2}$$

$$G_{1,3}(t) = \frac{1}{6}(3t^3 - 6t^2 + 4) \tag{3}$$

$$G_{2,3}(t) = \frac{1}{6}(-3t^3 + 3t^2 + 3t + 1) \tag{4}$$

$$G_{3,3}(t) = \frac{1}{6}t^3 \tag{5}$$

Therefore, the expansion of cubic spline can be written:

$$P_{i,3}(t) = \sum_{k=0}^{3} P_{i+k} G_{k,3}(t) = \mathrm{P}_i \cdot G_{0,3}(t) + \mathrm{P}_{i+1} \cdot G_{1,3}(t) + \mathrm{P}_{i+2} \cdot G_{2,3}(t) + \mathrm{P}_{i+3} \cdot G_{3,3}(t) \tag{6}$$

$i = 0, 1, 2, \ldots, m$。

From the above method, it can be deduced that when the $n \times m$ sub parametric surface is taken as the calculation surface, it represents $(n + 1) \times (m + 1)$ spatial points in space, which means the multiple polynomials composed of i and j are taken as the solution and solve it in the way of a matrix. The spatial grid formed by connecting the two adjacent points in the segment connection point set P_{ij} is used as the control network in turn.

$$P(u, v) = \sum_{i=0}^{n} \sum_{j=0}^{m} P_{ij} G_{in}(u) G_{jm}(v) \tag{7}$$

Where, $0 \leq u \leq 1, 0 \leq v \leq 1$, u and v represent parameters in two directions, $G_{in}(u)$ and $G_{jm}(v)$ represent spline basis functions in u direction and v direction in plane coordinates. And P_{ij} is the control vertex of the surface, $G_{in}(u)$ and $G_{jm}(v)$ are functions. Finally, the matrix of control point P_{ij} is obtained by the inverse calculation of the interpolation point [6].

According to the point cloud data of the tunnel, the surface reconstruction of the whole area is carried out by the cubic B-spline interpolation method based on MATLAB, and the surface results of the vault area are shown in Fig. 5.

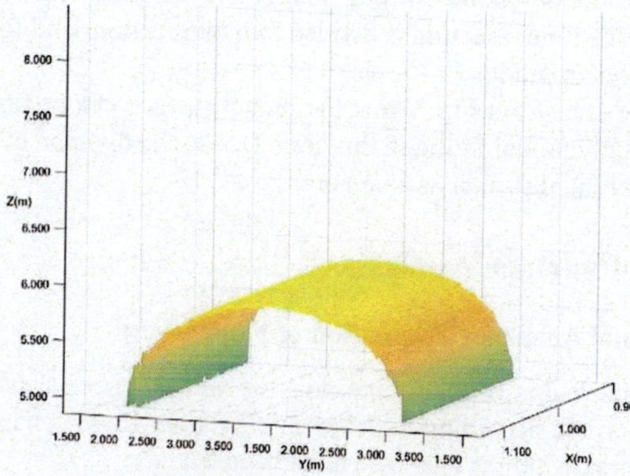

Fig. 5. Tunnel vault surface

The surface details are obtained by locally magnify Fig. 5, as shown in Fig. 6.

It can be seen from Fig. 6 that this method can restore the details of the tunnel vault completely, with high smoothness.

2.2.2 Construction of Camber in the Region of Arch and Waist

The curvature of the arch waist of the tunnel is small from the design drawing, which can be quickly calculated by low-order binary polynomials. In this paper, the point cloud data of 10m before and after the tunnel mileage K135,360 is extracted for local area surface reconstruction.

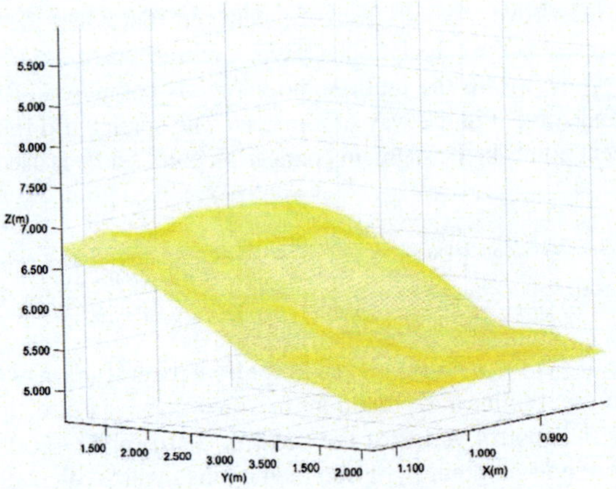

Fig. 6. Local enlargement of tunnel vault surface

In the point cloud data, the design axis is used as the X direction of the bottom surface, so the left and right sides of the arch waist need to be calculated separately. In this paper, the polynomial fitting method is used for surface reconstruction based on MATLAB. The tunnel section is divided into two sections and four areas for fitting. The results are shown in Fig. 7.

It can be seen from Fig. 7 that the tunnel surface can be obtained quickly through low degree polynomial fitting in this area. Due to the division of small areas, the overall effect of the tunnel surface is also good.

3 Tunnel Surface Analysis

3.1 Regional Accuracy Evaluation of Vault

By combining cubic B-spline interpolation and polynomial fitting, 3D reconstruction can be performed according to different curvature. In this chapter, the accuracy of the surface reconstruction is analyzed and evaluated.

First of all, the accuracy analysis of the vault area. In this paper, the accuracy of the 5 cm interval point cloud is evaluated by a 2 cm interval point cloud. The specific method is to randomly obtain 100 cloud data of 2 cm interval points in the selected area, a total of 4 groups, and record the corresponding z value of these points as z_{S}. Then the tunnel surface with a 5 cm interval point cloud is obtained by surface reconstruction, and the z value of the original point is recorded as z_{p}. Finally, the difference between the two z values is calculated as D, and the fitting error is calculated according to the difference. The calculation results are shown in Table 1.

$$\sigma = \sqrt{\frac{\mathrm{D}_1^2 + \mathrm{D}_2^2 \ldots\ldots + \mathrm{D}_N^2}{n}} = \sqrt{\frac{\sum \mathrm{D}_p^2}{n}} \tag{8}$$

According to Table 1, it can be seen that the reconstruction error of the 5 cm interval point cloud data is small and the fitting accuracy is high.

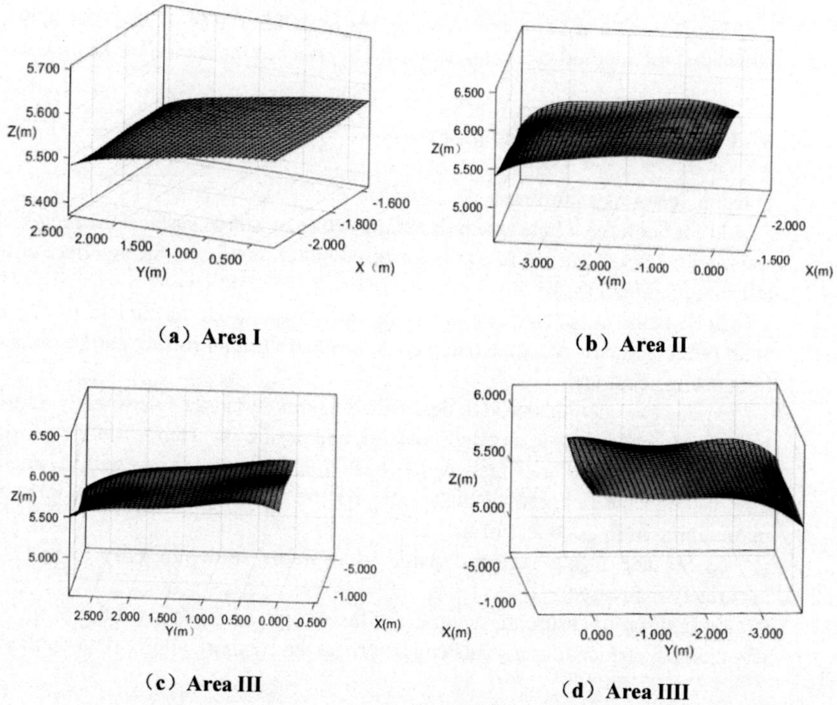

(a) Area I

(b) Area II

(c) Area III

(d) Area IIII

Fig. 7. Reconstruction of tunnel camber with two sections

Table 1. Fitting error of four groups of data

Group	No.1	No.2	No.3	No.4
Fitting error	1.003	1.020	1.005	0.990

3.2 Accuracy Evaluation of the Region of the Arch Waist

According to the polynomial fitting results of the arch waist region, not all point clouds are on the fitting surface, so the error of the data model is measured by calculating its goodness of fit R [10].

$$R = \frac{\sum\left(y - \hat{y}\right)}{\sum\left(y - \bar{y}\right)} \tag{9}$$

$\hat{y}$ represents the actual y value on the surface, and $\bar{y}$ represents the average y value of all points.

The goodness of fit of the four regions is 0.985, 0.979, 0.984, and 0.991, which shows that the method can get a better fitting effect in the region of the arch waist.

4 Conclusion and Prospect

The conclusion is as follows:

(1) Through the whole area polynomial fitting of tunnel surface, the region of tunnel waist with a small curvature can be restored better, while the fitting effect of the vault has some regional distortion.

(2) By using cubic B-spline interpolation to reconstruct the curved surface of tunnel vault (with large curvature), a complete and smooth tunnel surface can be obtained with high fitting accuracy.

(3) The two methods need to deal with the point cloud data pertinently, so that it can be arranged according to a certain interval, and ensure its accuracy in the area division.

(4) Fitting the tunnel arch waist area, that is, the position with a small curvature, through the polynomial fitting, can restore the tunnel surface completely, while maintaining high goodness of fit.

(5) Surface fitting region division needs more in-depth study to make it more accurately expressed.

(6) In order to ensure the accuracy of the fitting results, it is worthy of further study whether the surface fitting joints can be processed by more appropriate methods.

References

1. Nikoohemat, S., Diakitéb, A.A., Zlatanovab, S., et al.: Indoor 3D reconstruction from point clouds for optimal routing in complex buildings to support disaster management. Autom. Constr. **113**, 103109 (2020)
2. Li, Y.S., Jing, H.F.: Application of 3D laser scanning technology in wetland topographic map. Mapp. Spatial Geogr. Inf. **42**(12), 193–198 (2019)
3. Yao, X.X.: Research on Delaunay tetrahedralization of three -dimensional degenerate point sets. Tianjin University (2017)
4. Wu, X., Tang, N., Liu, B., et al.: A novel high precise laser 3D profile scanning method with flexible calibration. Optics Lasers Eng. **132**, 105938 (2020)
5. Duan, Q.C., Liu, T.: Experimental study on deformation monitoring technology of three-dimensional scanning in soft rock tunnel. Tunnel Constr. 1–8 (2020)
6. Zhang, B.: Research on image tracking algorithm based on particle filter. Shanghai Jiaotong University (2007)
7. Sun, S.J., Lin, H., Zheng, L.R.: Algorithm of adaptive NURBS interpolation points calculation. Minicomputer Syst. **35**(4), 895–899 (2014)
8. Su, J.K.: Studies on image interpolation based on the B-spline scheme. Guangdong University of Technology (2014)
9. Liu, S.Y., Han, Y., Jia, C.Q.: Cubic B-spline-interpolation-based mesh splicing and fusion. Chin. J. Meteorol. **23**(12), 1901–1909 (2018)
10. Zhou, P., Zhao, X.Z.: Optimized modeling of dynamic process oriented towards modeling error PDF shape and goodness of fit. Acta Automatica Sinica 1–10 (2020)

Author Index

J. J. Yang et al. (Eds.): GeoChina 2021, SUCI, p. 119, 2021.
https://doi.org/10.1007/978-3-030-79672-3